Language and Time

Language and Time

QUENTIN SMITH

New York   Oxford
OXFORD UNIVERSITY PRESS
1993

Oxford University Press

Oxford   New York   Toronto
Delhi   Bombay   Calcutta   Madras   Karachi
Kuala Lumpur   Singapore   Hong Kong   Tokyo
Nairobi   Dar es Salaam   Cape Town
Melbourne   Auckland   Madrid

and associated companies in
Berlin   Ibadan

Published by Oxford University Press, Inc.
200 Madison Avenue, New York, New York 10016

Oxford is a registered trademark of Oxford University Press

Library of Congress Cataloging-in-Publication Data
Smith, Quentin, 1952–
      Language and time / Quentin Smith.
         p.   cm.
      Includes bibliographical references and index.
      ISBN 0-19-508227-3
      1. Time.   2. Language and languages—Philosophy.   I. Title.
   BD638.S64   1993   115—dc20   92-37004

9 8 7 6 5 4 3 2 1

Printed in the United States of America
on acid-free paper

Preface

There are two theories of time: the *tenseless theory* and the *tensed theory*. The tenseless theory holds that temporal determinations consist only of the relations of *earlier than, later than,* and *simultaneous with*. The tensed theory of time (at least on one version) holds that temporal determinations also include the properties of pastness, presentness, and futurity. The tenseless and tensed theories also entail different theses about language, specifically about the semantic content of tensed sentences. The tensed theory implies, for example, that the indexical "now" in the sentence "The sun is now shining" conveys that the sun's shining has the property of presentness. But the tenseless theory holds that it does not and that "now" (as uttered on some occasion) merely refers to the date or time of the sun's shining or merely conveys that the sun's shining stands in some temporal relation (such as being simultaneous with the utterance of the sentence).

One of the aims of this work is to defend the tensed theory of time. This requires that the tensed theory be defended not only against the criticisms made by tenseless theorists such as D. H. Mellor and L. Nathan Oaklander, but also against the most prevalent theory in contemporary philosophy of language, the *New Theory of Reference,* which entails that temporal indexicals such as "now" directly refer to times and do not ascribe properties of presentness, pastness, or futurity. The defense of the tensed theory occupies Part I.

My argument for the tensed theory of time is a means to the end of establishing a more comprehensive theory about language and time. I shall call this theory *presentism*. According to this theory, every possibly true sentence includes presentness in its semantic content. This implies that all possibly true tenseless sentences, including all sentences that seem to have a timeless semantic content (such as "Two plus two equals four") include components that ascribe a presentness-involving property to something. I will argue that every possibly true sentence is synonymous with a presentist sentence, that is, with a sentence that has "presentness" for its grammatical subject and an expression of a form such as "inheres in such-and-such" for its grammatical predicate. For example, it will be maintained (despite initial appearances) that "Two plus two equals four" is synonymous with "Presentness inheres always in the equality of four to two plus two." This will require a development of a detailed theory of the nature of the *de re* or *de dicto* propositions expressed by sentences and of the states of affairs that correspond to true propositions.

The defense of presentism will also require a critical confrontation with Einstein's Special Theory of Relativity. It is perhaps implicitly obvious in the preceding remarks that presentism entails that presentness inheres in events or states *absolutely*, that is, that time is absolute, not relative. The argument for presentism will thus involve an argument that Einstein's Special Theory of Relativity does not characterize time. It will be shown that Einstein's theory entails merely that certain observable luminal relations are possessed relatively to a reference frame and that temporal relations and properties are possessed absolutely, even though in the case of distant physical events we may not be able to know their temporal relations or properties. The defense of the presentist theory of language and time will be the concern of part II.

I began work on this treatise in the fall of 1983 and the final draft was completed in the fall of 1990. I am especially indebted to William F. Vallicella for helpful critical comments on several of the successive drafts. William R. Carter also provided helpful suggestions about a penultimate draft that led me to delete long introductory and concluding parts and to add the chapter on Einstein's theory of relativity, as well as make smaller improvements. A major stimulus to the theories developed in this book is the critical comments offered by L. Nathan Oaklander on some of my published and unpublished articles on the tensed theory of time. Hector-Neri Castañeda, William Lane Craig, Storrs McCall, D. H. Mellor, Gilbert Plumer and J.J.C. Smart also provided comments in correspondence or discussion that proved useful in writing the final draft.

Yellow Springs, Ohio   Q. S.
February 1993

Contents

I

THE ARGUMENT FOR THE
TENSED THEORY OF TIME

1

The Translation Method and the Tensed and Tenseless Theories of Time

1.1 The Translation Method

The current opinion of proponents of the tenseless theory of time is that the argument that *tensed sentences (or sentence-utterances) are untranslatable by tenseless sentences (or sentence-utterances)* is insufficient to establish the tensed theory of time. Indeed, it is widely believed today that this argument provides no evidence at all for the tensed theory of time.[1] I shall make the case in this chapter that this belief is false. Specifically, I shall contend that an argument for the untranslatability thesis, along with a defense of the logical coherency and scientific viability of the tensed theory of time, provides adequate evidence for the tensed theory. After I make this case, I shall proceed to argue (in chapters 2–4) that tensed sentences and sentence-utterances *are* untranslatable by tenseless ones. This will prepare the way for the defense of presentism in part II.

Before I argue for the adequacy of the untranslatability thesis, it is necessary to explain what I mean by "translation" and "the translation method." The translation method involves the employment of the thesis

> (A) Two sentences as used in the same context have the same meaning if and only if they have the same truth and confirmation conditions and are logically equivalent and logically identical.

Thesis (A), which I shall explain shortly, may be applied to any pair of sentences to determine whether the two sentences, as used in the same context, translate each other. The application of (A) to a pair of sentences is the *translation method of linguistic analysis.* As I use the word *translate,* two sentences as used in the same context translate each other if and only if, in that context, they are synonymous, that is, have the same meaning or express the same proposition. My main employment of (A) in part I involves comparing certain logically contingent tensed sentences with certain logically contingent tenseless sentences and arguing that they do not translate one another. For example, I shall argue that "The sun is now shining" as

3

said at noon, September 28, is not translated by "The sun is shining at noon, September 28," where the "is" is tenseless.

Thesis (A) should be understood as stating three necessary conditions of translation and one condition that is both necessary and sufficient. Regarding the three necessary conditions, it may be said that a sentence S_1 as used in a context C is synonymous with, or translated by, a sentence S_2 as used in C only if

(1) S_1 and S_2 as used in C have the same truth conditions.

(2) S_1 and S_2 as used in C are confirmed or disconfirmed to the same degree by the same observations.

(3) S_1 and S_2 as used in C are logically equivalent, that is, entail each other.

Sameness of truth conditions and confirmation conditions are each insufficient for translatability if only for the reason that two nonsynonymous analytic sentences are true under all conditions and stand in the same relation to all possible observations. Logical equivalence is also insufficient, since "The sun has a shape" is logically equivalent to "The sun has a size," yet these two sentences are synonymous only if the two adjectives "size" and "shape" are synonymous, which they are not. Moreover, conditions (1)–(3) are not jointly sufficient for synonymy, since "The sun has a shape" and "The sun has a size" as used in C meet all three conditions yet are nonsynonymous.

A condition of translatability that is both necessary and sufficient is that S_1 and S_2 as used in C are *logically identical*, that is, that

(4) S_1 and S_2 as used in C refer in the same way(s) to the same item(s) and ascribe to said item(s) the same monadic or polyadic property or properties.

Logical identity, unlike conditions 1, 2, and 3, is unfamiliar and requires some words of explanation, particularly regarding the phrase "refer in the same way(s) to the same item(s)." Referring to something *directly* is a way of referring to something; and there are many ways of indirectly referring to something, one for each different descriptive sense that a referring expression may express. If the definite descriptions are used attributively, "The morning star is a planet" and "The evening star is a planet" both refer to the same item, Venus, but in different ways—the former, in respect of its property of being visible to Earth in the morning, the latter, in respect of its being visible in the evening. Likewise, both refer to Venus in a different way from "Venus is bright," supposing that "Venus" is used in this sentence as a directly referring name. (By "a sentence refers" I mean "a part of the sentence refers," and by "a sentence ascribes a property" I mean "a part of a sentence ascribes a property.")

Examples of translation in my sense are the translation of "Jane departed from the unsafe trail" by "Jane left the dangerous path" as used in the same context, the translation of "Cars are used by humans as instruments" by "Automobiles are employed by human beings as tools" and the translation of "Some feline creature died on May 1, 1865" by "At least one cat expired on the first day of May in 1865."

There are many questions that may be raised about this definition of translation. For example, the reader may wonder why I need to introduce the merely necessary conditions of translation (1)–(3) if (4) is both necessary and sufficient. The reason is that I wish to accommodate objections to the effect that (4) is an implausible or too-strict condition of translatability. My application of translatability thesis (A) to tensed and tenseless sentences will involve showing that tenseless sentences not only fail to meet the condition of being logically identical with tensed sentences (as used in a context C) but also fail to meet the three less stringent necessary conditions of being translations of the tensed sentences (as used in a context C). Thus, my arguments will suffice to refute those who claim that the relevant tenseless and tensed sentences express the same propositions since they meet the condition of logical equivalence or sameness-of-truth conditions or sameness-of-verification conditions.

It may also be wondered whether some of my conditions of translatability are redundant. For example, does not "S_1 and S_2 have the same truth conditions" mean the same as "S_1 and S_2 are logically equivalent"? I believe the answer is negative; for "the truth conditions of S_1 and S_2" does not refer to a relation between S_1 and S_2 but to the states of affairs that make S_1 and S_2 true, whereas "the logical equivalence of S_1 and S_2" refers, instead, to a relation (i.e., mutual entailment) between S_1 and S_2. Perhaps it is true that "S_1 and S_2 have the same truth conditions" is logically equivalent to "S_1 and S_2 are logically equivalent"; but logical equivalence is not logical identity, and the fact that the relation of sameness-of-truth conditions is not *identical* with the relation of logical equivalence justifies the claim that these are different conditions of translatability. This distinction between the two translation conditions is not otiose, for it will prove useful in the arguments I shall develop in chapters 2–4.

It may be objected at this point that all four of my conditions of translation are too strong and that the tenseless theory of time can be established by using some condition weaker than all of them, such as the *intersubstitutivity salva veritate* in extensional contexts of the relevant tensed and tenseless sentences. The detenser might argue that the fact that "The meeting starts on March 26, 1989" is inter-substitutable *salva veritate* with any March 26, 1989 extensional occurrence of "The meeting is starting now" vindicates the tenseless theory. My response to this objection is that I will show that the failure of tensed and tenseless sentences to meet each of conditions (1)–(4) provides all the justification that is needed for the success of the semantic argument for the tensed theory of time. Even if these sentences meet this extensional condition of translation, they do not meet my four conditions; and the latter fact is all that I shall require to argue that the tensed sentences express propositions that ascribe presentness or a property that includes presentness.

My introduction of the three necessary conditions (1)–(3) also suffices to meet the objection that "tenseless sentences analyze tensed sentences even if they do not translate them." This objection is based on the true belief that *translation* is a different relation than *analysis* since the former is symmetrical but the latter asymmetrical; "Socrates is a rational animal" is an *analysans* of "Socrates is a human," but the latter is not an *analysans* of the former. It would seem to follow that "Tenseless sentences do not translate tensed sentences" does not entail that "Tenseless sentences do not analyze tensed sentences" and thus that my semantic argument for the tensed theory is jeopardized. However, the fact that I introduce conditions (1)–(3) meets this objection, since at least one of these conditions (e.g., logical equivalence or sameness-of-truth conditions) is also a necessary condition of analysis on any standard definition of "analysis."

There are other objections to my use of the method of translation that may be made by proponents of the four main versions of the tenseless theory of time, but these objections are of such a serious nature that I shall need separate sections to deal with them (see 1.2–5). In the remainder of the present section I shall concentrate on explaining my employment of this method more precisely by offering definitions of the sorts of tensed and tenseless sentences that I shall investigate.

I shall use translation thesis (A) in chapters 2–4 by applying it to certain classes of tensed and tenseless sentences, which I shall call (loosely following the terminology, if not strictly the ideas, McTaggart introduced) *A-sentences* and *B-sentences*. In order to define these sentences, I need the terms *A-property, B-relation,* and *B-position*. An A-property (or A-position) is a monadic property of *presentness, present pastness,* or *present futurity*. (For brevity's sake I shall usually omit the qualifying "present" in mentioning the latter two A-properties.) Pastness and futurity, of course, admit of subdivisions (e.g., being past by two minutes). Any property partly composed of presentness, apart from the two properties of pastness and futurity, is not an A-property. For example, *being always present* is not an A-property. One reason for this terminological decision is that I shall define B-sentences, in part, as those that do not ascribe A-properties; and I shall argue in chapter 6 that B-sentences nonetheless ascribe some complex property of which presentness is a part. B-relations are *earlier, later* and *simultaneous;* and B-positions are dates defined in terms of these relations, such as 10 B.C., (defined as the time that is ten years earlier than Christ's birth).

An A-sentence is defined in terms of five characteristics:

  i. It contains a future, present, or past tensed copula (e.g., "The sky *was* red") and/or a future, present, or past tensed verb ("He *ran*"). It may also contain a temporal indexical, such as a temporal adverb ("The meeting starts *now*") or temporal pronoun ("*Yesterday* was a sunny day.").
 ii. It is explicitly or implicitly about an event. "The collapse of the coliseum is imminent" is explicitly about an event, since its grammatical subject refers to an event. "The coliseum is about to collapse" is only implicitly about an event, since its grammatical subject refers to a thing; this sentence implies that there will be an event (the collapse) involving the thing the sentence is about (the coliseum). I use the word *event* in a very broad sense, so that any inherence of any property in something counts as an event; thus, the inherence of collapsing in the coliseum and the inherence of redness in the sky are both events.[2]
iii. It refers to the event's temporal position or ascribes to the event an n-adic temporal property. (I leave undecided at this juncture whether the temporal position is an A-position or B-position and whether the n-adic property ascribed is a monadic A-property or a polyadic B-property. I also leave

undecided whether the reference to the temporal position is direct or indirect (i.e., mediated by a sense).

iv. It refers to or ascribes a temporal position or property in such a way that the position or property to which it relates is determined by its tense. A past tensed sentence ascribes either the A-property of pastness or the B-property of being earlier or refers to a B-position that is earlier than the time of use of the sentence, and the same holds mutatis mutandis for a present and a future tensed sentence.[3]

v. It is logically contingent but is neither a universal generalization, (like "Any past crow was black, any present crow is black, and any future crow will be black") nor an omnitemporal disjunction (like "The death of the sun has occurred, is occurring, or will occur").

Any sentence with all five of these characteristics is an "A-sentence." An A-theorist, or tenser, is a philosopher who believes that these sentences relate to A-positions; and a B-theorist, or detenser, is a philosopher who believes that they do not. (Of course they are some philosophers, such as A. N. Prior, F. Christensen, G. Lloyd. G. Plumer, A. B. Levison, and E. Lowe[4] who hold that A-sentences relate to "A-positions" in some sense but deny both that A-positions are *monadic properties* and that they are individuals or relations. But these philosophers have never succeeded in explaining what their unusual A-positions are; and their doctrine is ultimately incoherent, as I shall argue in part II.) A detenser also believes either that A-sentences or their tokens are "translatable" in some sense by B-sentences or their tokens—or at least that the A-sentences entail nothing true about the nature of time that is not entailed by B-sentences. A B-sentence has four characteristics:

i. It does not contain a tensed copula or verb or a temporal indexical unless, perhaps, within a subclause that is in quotation marks.

ii. It is explicitly or implicitly about an event.

iii. It refers directly or indirectly to the event's B-position or ascribes to the event a polyadic B-property but does not ascribe to it an A-property.

iv. It is logically contingent but is not a universal generalization.

If we place parentheses around the relevant copulae and verbs to indicate that they are tenseless, we can give as examples of B-sentences: "He (is) running later than this utterance" and "There (is) somebody who (is) born on May 1, 1960." I would note that these four characteristics do not entail that B-sentences do not ascribe a property of which presentness is a part, since there are some properties of which presentness is a part that are not A-properties. For example, *being either past, present, or future* and *being everlastingly present* and conditions i–iv are consistent with one of these properties' being ascribed by B-sentences. However, I shall not assume this at the outset, since that would obviously be question-begging. In this and the next three chapters I shall assume only that B-sentences relate to B-positions or relations and not until chapter 6 shall I argue that they ascribe one of these complex properties.

My employment of the translation method to establish the tensed theory of time can now be stated more precisely. In chapters 2 and 3, I argue that no A-sentence or

A-sentence-token is translatable by any B-sentence or B-sentence-token; and in chapter 4, I argue that the best explanation of this fact is that the A-sentences ascribe A-properties. Given the assumption that some A-sentence-tokens (e.g., "The sun is now shining") are true, it follows that the tensed theory of time is true, that is, that A-properties are in fact exemplified.

I turn now to the major objections to my translation method that might be made by proponents of the tenseless theory of time. I distinguish four versions of the tenseless theory of time, each of which is discussed in a separate section. In section 1.2, I discuss the *old* tenseless theory of time. (This and the other names of the versions of the tenseless theory are mine.) In section 1.3, I discuss the *new* tenseless theory of time. In section 1.4, I discuss the a priori version of the *nonsemantic* tenseless theory of time; and in section 5, the a posteriori version of the same theory.

1.2 The Old Tenseless Theory of Time

What I have elsewhere called the old tenseless theory of time[5] was initiated by Russell in 1903 and is associated with such detensers as A. J. Ayer, Milton Fisk, Nelson Goodman, W.V.O. Quine, J.J.C. Smart (in his early period), Clifford Williams, and others.[6] This theory holds that A-sentences are "translatable" in some sense by B-sentences and that A-sentences consequently do not ascribe A-properties but merely B-relations. My use of the translation method would beg the question against the old tenseless theory if I began with a definition of A-sentences that stated they expressed A-propositions and defined the latter as propositions that ascribe A-properties to events, for these definitions would rule out in advance the possibility that A-sentences are translated by B-sentences. But I think it is obvious that my translation method does not beg the question against the old tenseless theory, since "expressing propositions that ascribe A-properties" or "ascribing A-properties" are not among my five defining characteristics of A-sentences and no characteristic is included that is logically inconsistent with the old tenseless theory of time. My translation method avoids question-begging assumptions, since it is a means of investigating *sentences;* and theses about the specific nature of the *propositions* expressed by these sentences or the *temporal positions* to which they relate are not assumptions of, but conclusions reached by, the employment of this method.

It might seem that the tenser can easily formulate premises that do not beg the question against the old tenseless theory of time, but the history of the debate between tensers and the old detensers shows that such premises are not as easy to come by as one might think. In fact, the most detailed and systematic semantic defense of the tensed theory against the old tenseless theory of time that has been offered to date, Richard Gale's defense in *The Language of Time,* is based on question-begging premises. An examination of Gale's argument will make clear precisely the sort of premises that we must avoid and the type of argumentation that is required if the old tenseless theory of time is to be soundly refuted. (Gale has since changed his allegiance and now accepts Michelle Beer's version of the new tenseless theory of time, which I shall argue to be unsound in chapter 2.)

In his *Language of Time,* Gale begins by defining A-sentences as sentences that

express A-propositions (which he calls A-statements) and by defining A-propositions as possessing among their necessary characteristics a reference to an event
with an A-property: "An A-statement asserts that an event has a given A-determination."[7] It is to be noted that Gale is not here *arguing* that A-sentences express
A-statements, or that A-statements assert that events have A-properties; rather, he is
defining them to do so. It is on the basis of these definitions that Gale *then* argues
that the old tenseless theory of time is false. In the following passage, Gale is
considering, among other claims, the claim that the A-statement expressed by "S is
now φ" is translated by the B-statement expressed by "S's being φ (is) simultaneous
with theta," where "theta" is a metalinguistic proper name for the occurrence of the
tensed token in the *analysandum:*

> The B-statements in the *analysans* of these two analyses do not entail the A-state-
> ment in the *analysandum:* that S's being φ *is* simultaneous with *theta* (the occur-
> rence of a token of "S is now φ") does not entail that S is now φ. These B-state-
> ments describe a B-relation between S's being φ and a certain token event without
> entailing that either one of these events is now present (past, future). That they do
> not convey or entail information about the A-determination of an event can be seen
> by the fact that whenever someone uses the sentence, "S's being φ *is* simultaneous
> with *theta* (the occurrence of a token of 'S is now φ')," he has not forestalled the
> question whether S's being φ (or the occurrence of *theta*) is now present (past,
> future).[8]

In this argument, Gale is assuming precisely what the defender of the old tenseless
theory contests, namely, that events have A-determinations and that the statements
expressed by A-sentences convey information about events' A-determinations. His
argument has this structure:

(1) A-sentences express A-statements.

(2) A-statements convey information about the A-determinations of events.

(3) B-statements convey information about the B-determinations of events but
not about their A-determinations.

(4) Therefore, B-statements cannot translate A-statements.

Gale did raise the issue of whether (2) is question-begging, but he did not state or
resolve it satisfactorily. He reasoned that if B-statements translate A-statements,
they must ascribe A-determinations and, consequently, that if ascriptions of A-determinations were a sufficient condition of a statement's being an A-statement, then the
B-statement would really be an A-statement and the detenser would not have succeeded in showing that B-statements translate A-statements. To avoid begging this
issue against the detenser, Gale stipulates that ascription of A-determinations is a
necessary but not sufficient condition of an A-statement.[9]

But this is to misidentify the real respect in which (2) begs the issue, for the old
tenseless thesis is not that B-statements translate A-statements regardless of whether
it be true that both these statements ascribe A-determinations or that both fail to
ascribe them but that B-statements do not ascribe A-determinations and do translate
A-statements, which therefore also do not ascribe A-determinations. Consequently,

to stipulate that a necessary condition of an A-statement is that it ascribe A-determinations is to assume, rather than prove, the old tenseless thesis to be false.

To refute the old tenseless thesis in a non-question-begging way, (2) must be placed among the conclusions, rather than the premises of the argument. Item (1) can be allowed as a premise if A-propositions are not defined as propositions that ascribe A-determinations but are defined solely as the propositions expressed by A-sentences. Whether these propositions ascribe A-determinations, rather than B-determinations, is a matter to be established by argument.

The underlying problem with Gale's argument against the old tenseless theory of time is that he did not perform a "semantic ascent" at the point where this ascent was needed. Instead of beginning with A-sentences and their rules of use and arguing from these rules to conclusions about the propositions and temporal positions to which these sentences relate, he began by assuming things about the very propositions and positions that were in dispute.[10] The lesson to be learned is that a sound refutation of the old tenseless theory of time requires the tenser to formulate premises about the data that he and his opponents agree in advance to exist. Both parties agree that there are A-sentences and B-sentences and that there are conventionally accepted rules of usage governing these sentences. For example, both agree that there are commonly accepted rules governing the truth conditions and entailment relations of these sentences. Moreover, both agree that these sentences have rules of usage that are statable in a *tenseless* language. The tenser would beg the question if he assumed from the outset that A-sentences *also* have irreducibly tensed truth conditions, confirmation conditions, and so on that imply that events have A-properties; and the detenser would beg the question if he assumed from the outset that A-sentences do *not* also have such conditions. But since they both agree that A-sentences have tenseless rules of usage, the explication and discussion of these rules forms a common ground from which both the tenser and detenser can build their arguments.

The restriction of the premises to tenseless rules of usage is not unfair to the tenser because the tenser acknowledges that events are B-related and that A-sentences imply facts about their B-relations. Furthermore, the tenser acknowledges that their B-relations are describable in a tenseless language. The thesis the tenser wants to establish is that A-sentences *also* convey information about the A-properties of events and that the rules governing the conveyance of this information are irreducibly tensed. The detenser, on the other hand, is in a different position. The detenser denies that events have A-properties and therefore denies that there are irreducibly tensed rules governing the ascription of these properties. Consequently, for the tenser to begin by assuming that there are such rules is to assume from the outset that the detenser's thesis is false, a practice that is not going to convince any detenser. The tenser must, instead, *argue* that there are such rules, just as the detenser must *argue* that there are no such rules. The common ground they have is the assumption that A-sentences have tenseless rules of usage; it is from this assumption that both of their arguments should begin.

I shall conduct my arguments in chapters 2–3 in terms of these tenseless rules of usage. I will show that A-sentences have different tenselessly describable truth conditions, confirmation conditions, entailment relations, and logical identities than

B-sentences and are therefore untranslatable by them. I shall then argue (in chapter 4) that these tenseless rules of A-sentences require that A-sentences also have tensed rules and that the latter rules imply that A-sentences express A-propositions about events with A-properties.

It remains to show that my argumentative strategy can be soundly employed not only against the old tenseless theory of time but also against the new tenseless theory of time and the two nonsemantic tenseless theories.

1.3 The New Tenseless Theory of Time

The difference between the new and old tenseless theories of time (as defined elsewhere),[11] pertain to the relevance of synonymy or translatability. The defenders of the old tenseless theory argue that A-sentences as used on any given occasion are translatable in some sense by B-sentences and therefore that A-sentences do not ascribe A-properties. The proponents of the new theory, on the other hand, hold that tokens of A-sentences are not translatable by B-sentences but that since B-sentences suffice to give the truth conditions of A-sentences or their tokens, it follows that A-sentences do not ascribe A-properties. Although intimations of the new tenseless theory of time appeared as early as 1973 with Stephen Braude's "Tensed Sentences and Free Repeatability," the new tenseless theory of time was inaugurated by J.J.C. Smart in his 1980 essay "Time and Becoming" and subsequently developed (in one version or another) by D. H. Mellor, Murray MacBeath, Paul Fitzgerald, Michelle Beer, Richard Gale, and others.[12] Smart writes in "Time and Becoming" that Richard Taylor (a tenser) argued that the tenseless theory of time is false, since A-sentences are not translatable by B-sentences but that Taylor's argument is invalid since all the detenser need show is that A-sentences have truth conditions statable entirely in the tenseless metalanguage of the B-theory:

> Taylor's criticism of "the attempts to expurgate becoming" seem to me to turn on the impossibility of translating expressions, such as tenses, into non-indexical ones. I agree on the impossibility, but I challenge its metaphysical significance, since the semantics of indexical expressions can be expressed in a tenseless meta-language. . . . [For example,] when a person P utters at a time t the sentence "Event E is present" his assertion is true if and only if E is at t.[13]

Thus, even if "E is at t" does not translate the utterance of "Event E is present" that occurs at t, the fact that it gives the truth conditions of the latter shows that the temporal conditions necessary and sufficient for the A-token's truth are B-conditions and therefore that the postulation of A-conditions in addition are superfluous and unjustified. An especially vivid statement of this central thesis of the new tenseless theory of time is given by Paul Fitzgerald, who writes of D. H. Mellor's *Real Time* that it

> concedes to Gale [in his tenser period] and other objectivist A-theorists (believers in the consciousness-independent reality of A-determinations) their claims that A-statements cannot be translated into non-A-statements [but also shows that]

A-determinations are not needed to give an account of the truth-conditions of tensed statements. So A-determinations are otiose.[14]

It would seem at first glance that proponents of this new tenseless theory of time would not be in the least bit troubled by my argument in part 1, since my argument merely establishes that A-sentences and their tokens are untranslatable by B-sentences and their tokens, a point already conceded by these proponents. The "new detensers" would remark that my method of translation is inadequate to establish the tensed theory of time and therefore that my project is doomed from the start.

My response to this challenge is that my method of translation will be used to establish more than that A-sentences and their tokens are untranslatable by B-sentences and their tokens; it will also be used to establish that the truth conditions of A-sentences and tokens cannot be adequately stated by B-sentences. It will be shown that B-sentences such as "E is at t" either do not state any of the truth conditions of A-sentence-tokens (e.g., the token of "E is present" that occurs at t) or do not state all of their truth conditions and that A-sentences must be used for an adequate statement of these conditions. It will be argued, furthermore, that this fact (along with others) entails that tokens of A-sentences ascribe A-properties. This can be established by the translation method, since this method involves a reference to the truth conditions of sentences and thereby enables relevant conclusions about these truth conditions to be drawn. Specifically, this method will be employed to show that no A-sentence or A-sentence-token has the same truth conditions as any B-sentence or B-sentence-token and that this implies that the new tenseless theory of time is false. To see that this implication in fact holds, consider a B-sentence (type or token) such as "E is at t" used to state the truth conditions of an A-sentence-token such as "E is present." Now clearly, it is true that

(1) "E is at t" is true if and only if E is at t.

But if no A-sentence-token has all and only the same truth conditions as any B-sentence (type or token), then the token of "E is present" that occurs at t cannot have all and only the same truth conditions as "E is at t" (understood as type or token); thereby, it is false that

(2) "E is present" as uttered at t is true if and only if E is at t

gives all the truth conditions of "E is present" as uttered at t. This will be shown in detail in the following chapters; I mention it here only to show that it is not obvious at the outset that the new tenseless theory of time shows my translation method to be unviable.

L. Nathan Oaklander believes that this way of arguing against the new tenseless theory of time is misdirected, since it is based on a misrepresentation of this theory. He has replied to my earlier criticism of the new tenseless theory of time by claiming that my arguments do not count against the new tenseless theory, since this theory does not make the claims I attributed to it.[15] He concedes my thesis that the new tenseless theory of time fails to provide truth conditions that are adequate to the meaning of ordinary A-sentences but claims that this is irrelevant, since the new tenseless theory does not aim to give these truth conditions. This theory, according to Oaklander, does not aim to explain the meaning of ordinary sentences but to develop a semantics that is adequate to the metaphysical nature of time. In his terminology, the new tenseless theory aims to construct not a language that has the "logical" function of representing the meaning of ordinary language but, instead one that merely has the "ontological" function of representing the metaphysical nature of time:

> The new tenseless theory accepts the tenser's claim that tensed discourse and thought are ineliminable. It therefore agrees that any logically adequate representation of temporal language, that is, any language capable of representing the meaning and logical implications of our ordinary talk about time, must be tensed. The detenser denies, however, that from an ontological point of view, a perspective that attempts to represent the nature of time, that tense is ineliminable. Smith understands very well that recent detensers maintain that tensed sentences cannot be replaced by tenseless ones without loss of meaning. What he fails to appreciate is that in accepting the irreducible nature of tensed discourse, the new tenseless theory is abandoning the analytic ideal of arriving at a single language that is adequate for both ontological and logical investigations. Once these two functions of language are separated and kept distinct, it is open to the defender of the tenseless view to maintain that logical connections among sentences in ordinary language do not represent ontological connections between facts in the world. [16]

It seems to me, however, that it is Oaklander who misunderstands the new tenseless theory of time. Oaklander confuses this theory with the nonsemantic tenseless theory of time, which I discuss in the following two sections. The new tenseless theory of time, as defined in my article[17] and as espoused by Smart, Mellor, MacBeath, and others, is based on the thesis that the tensed theory of time is false *on the grounds that the truth conditions of ordinary A-sentences and their tokens can be stated in a tenseless metalanguage;* that is, it is the theory that tenseless truth condition sentences provide a "logically adequate representation of ordinary temporal language" and therefore that the tenseless theory of time is true. This is precisely the point of Smart's claim that "the semantics of indexical expressions can be expressed in a tenseless metalanguage."[18] Smart is here saying that the tenseless metalanguage adequately expresses the meaning of (in the sense of "gives the truth conditions of") ordinary indexical expressions such as "E is present." Smart infers from this the thesis, of "metaphysical significance,"[19] that the tenseless theory of time is true. Contra Oaklander, Smart is concerned with the logical structure of ordinary language and is interested in how the meaning of ordinary expressions should be understood or represented in theories of meaning for ordinary language. Thus, he makes such remarks as: "I think that ordinary adverbs should be understood in terms of predicates of events, as has been suggested by Donald Davidson. Tenses should be handled differently, by means of a tenseless metalanguage."[20] Clearly, "tenses" here means "ordinary tenses" and "handled" means "treated in a way that perspicuously reveals their semantic content."

This is even more clearly the case in D. H. Mellor's theory, which is the most developed version of the new tenseless theory of time. Mellor explicitly and repeatedly says that his representation of the truth conditions of ordinary A-sentences and their tokens captures the semantic content or meaning of these tokens, for example, when he says that "anyone who knows that for any place X tokens of 'X is here' are true if and only if they are at X, and that for any date T tokens of 'It is now T' are true if and only if they occur at T, surely knows what '. . . is here' and 'It is now . . .' means in English"[21]—a comment Mellor reproduces almost verbatum in another context on the next page. For Mellor, tensed sentences "may not *have* the same meaning as the tenseless sentences that give their truth conditions, but those truth conditions surely *give* their meaning."[22] Mellor's truth condition theory aims to explain correct ordinary usage of tensed sentences: "Correct usage *is* explained by people knowing how the truth of what they say depends on when and where they say it; in particular, the *different* meanings of different sentences are differentiated, as they are not in mathematics, by their different truth conditions."[23] Again, "The token-reflexive truth conditions of tensed sentences incidentally explain not only how we usually use them, but also how our usage varies when their tokens take time to arrive."[24]

Thus, Oaklander is wrong when he describes the new tenseless theory of time of Smart and Mellor as not having the aim of constructing tenseless truth conditions sentences constitutive of "a logically adequate language—a language that represents [in truth condition sentences] the meanings and entailments of sentences in a natural language."[25] But this fact does not eliminate the relevance of Oaklander's article for determining the soundness or unsoundness of my translation method, for Oaklander's theory can be abstracted from its context as an alleged defense of the new tenseless theory of time and considered on its own merits as a version of the *nonsemantic tenseless theory of time*.

1.4 The A Priori Version of the Nonsemantic Tenseless Theory of Time

Oaklander's tenseless theory of time is nonsemantic in the sense that it considers the meaning or semantic content of ordinary tensed discourse as irrelevant—or at least as not crucial to the truth or falsity of the tenseless theory of time. There is a distinction between the "ontological language," which represents the real nature of time and "ordinary language," which does not. The best way to approach the nonsemantic theory is to ask for the justification of its thesis that ordinary temporal discourse does not represent the real nature of time. It perhaps goes without saying that one cannot appeal to the thesis that the following argument form is valid:

(1) Aspect A of ordinary language has the semantic property F.

Therefore,

(2) Reality contains no part or property that is represented by A's F-ness.

The sequence (1)–(2) is an obviously invalid argument form; for if it were valid, the following argument would be sound:

(3) In ordinary language, any sentence of the form "x has a shape" entails one of the form "x is spatially extended."

Therefore,

(4) It is false that in reality something exemplifies *having a shape* only if it exemplifies *being spatially extended.*

Clearly, the nonsemantic tenseless theory must be understood as endorsing the weaker claim that one cannot be justified in believing a sentence of the form of (2) or of the negation of (2) on the basis of a sentence of the form of (1). But what justifies this weaker claim?

There are two possible ways to justify this claim, on the basis of a posteriori considerations or on the basis of a priori considerations. The a posteriori way gives us the a posteriori nonsemantic tenseless theory of time. This theory holds that one is not justified in believing the ordinary tensed representation of time, since there are empirical reasons, based on current scientific theory, for believing that the ordinary tensed representation is false. This a posteriori theory, held by such detensers as Adolf Grünbaum, will be discussed in the next section.

The predominant a priori version of the nonsemantic theory implies that we are not justified in accepting the ordinary tensed representation of time, since there are logical reasons for thinking this representation false. These logical reasons are usually subsumed under the heading of "McTaggart's paradox." This version of the nonsemantic tenseless theory is the one defended by Oaklander.[26] I shall discuss McTaggart's paradox in chapter 5 and show that it provides no reason for rejecting the ordinary tensed representation of temporal reality.

However, there is a second possible form of the a priori nonsemantic tenseless theory; and this form will be the topic of the rest of the present section. The second form denies that the translation method can establish the tenser's thesis *even if* the tenser counters the scientific and McTaggart-based arguments. The first form of the a priori theory (the "logical reason" form) held that there are *logical reasons,* based on McTaggart's paradox, for rejecting the ordinary tensed representation of time. But the second form, which I shall call the "skeptical nonsemantic theory," holds that even in the absence of logical or empirical reasons for rejecting the ordinary representation of time, there is no reason to accept it. According to the skeptical nonsemantic theory, an inference from

(1) Aspect A of ordinary language has the semantic property F

to

(5) It is more reasonable to believe, than not believe, that reality has a component that is represented by A's F-ness.

is invalid even if there are no empirical or logical reasons to disbelieve that reality has a component that is represented by A's F-ness. More specifically, this skeptical detenser claims that

(6) Ordinary A-sentences ascribe A-properties, some tokens of these sentences are ordinarily accepted as true, and there are no logical or scientific reasons to reject all these sentences as false

does not render it more reasonable to believe, than to disbelieve or remain neutral toward

(7) Some tokens of ordinary A-sentences are true, and events possess A-properties.

It seems to me, however, that this form of the a priori nonsemantic theory can be rejected out of hand, since it is committed to a self-referentially unjustifiable skepticism. If sentences of the logical form of (6) do not make it more reasonable to believe, than to disbelieve, or remain neutral toward sentences of the logical form of (7), then, by application of the same principle, it is not more reasonable for the skeptic to believe, rather than not believe, that (6) does not render it more reasonable to believe, rather than not believe, (7). This can be shown most clearly in terms of an example. This a valid argument:

(8) "E is present" is ordinarily accepted as true in circumstances that ordinarily seem to be those in which E is present.

(9) There are no logical or empirical reasons to not believe that "E is present" is true in the circumstances in which it is ordinarily accepted as true.

(10) There are no other considerations relevant to determining the truth or falsity of "E is present."

Therefore,

(11) It is more reasonable to believe that "E is present" is true in the circumstances in which it is ordinarily accepted as true than to not believe this.

Now our skeptic regards (8)–(11) as invalid since he holds that (8)–(10) are insufficient to justify accepting "E is present" as true. He allows that they are necessary conditions of the truth of (11) but that they are singly and jointly insufficient for its truth. Call the skeptical claim that "(11) is not entailed by (8)–(10)" the claim C. Is the skeptic, by his own standards, justified in believing C? To see that he is not, consider that his acceptance of C is purportedly justified by the principle p:

(p) If something S seems to be true, then, even in the absence of empirical and logical reasons to not believe S and in the absence of any other relevant considerations, S's seeming to be true does not justify a belief in S.

A belief B is justified by something J if J makes it more reasonable to believe B than to not-believe B. But what could possibly justify a belief in p? Suppose we grant for the sake of argument that there are no empirical or logical reasons to not-believe p. We cannot then appeal to the facts that p seems true to the skeptic and that there are no other relevant considerations, for p itself rules out this purported justification. This leaves us with "other relevant considerations" as the only possible justifications. Call any consideration that may be taken as a reason for believing p "R." What justification could there be for believing that R *is* a reason for believing p? By p, this justification cannot be that R seems to the skeptic to be a reason for believing p and that there are no other relevant considerations. Consequently, if the skeptic is to claim that his belief that R is a reason for believing p is justified, he must claim that there is some further consideration R′ that constitutes this justification. But if he is to maintain that this latter claim is justified, he must (by p) appeal to still further considerations R″. Clearly, an infinite regress ensues. This entails that the skeptic cannot be justified in believing p. This entailment goes through, since the skeptic is a finite mind and cannot comprehend an infinite number of steps in a chain of alleged justifications. But even if he could, his belief in p would still be unjustified, since he would have no justification for believing the proposition *that the infinite chain justifies belief in p,* since in this case, there would be no other relevant considerations; and this fact, along with the fact that this proposition seems true to him, would not, by p, justify his belief in this proposition.

It follows, therefore, that the rejection of the argument form instantiated by (8)–(11) cannot be justified by the standards of justification that are entailed by the rejection of this argument form and therefore that its rejection is self-referentially unjustifiable. By contrast, the acceptance of this argument form as valid is self-referentially justifiable, since if this argument form is valid, one can construct the sound argument "The argument form seems to be valid, there are no logical or empirical reasons to think it invalid, and there are no other relevant considerations, so it is justified to belief that it is valid." In other words, this argument form licenses self-justifying beliefs, one of which is the belief that the argument form is valid. A belief in something S is self-justifying if and only if nothing other than the belief in S, the absence of logical–empirical reasons for not believing S, and the absence of any other relevant considerations suffice to justify the belief in S.[27]

It might be objected by the skeptical detenser that there is a reason to think that this argument form is invalid, namely, that it has substitution instances that are invalid. It might be alleged that the following argument is clearly invalid:

(12) Ordinary religious-discourse sentences ascribe virtue properties to God, and such sentences are ordinarily accepted as true.

(13) There are no logical or empirical reasons to not-believe that God has virtue properties.

(14) There are no other relevant considerations to the truth or falsity of "God has virtue properties."

Therefore,

(15) It is more reasonable to believe, than not to believe, that God has virtue properties.

It might be said that this argument is invalid on the grounds that ordinary speakers may have no real basis or reason for ascribing virtue properties to God. They may just believe this, without any justification whatsoever.

I would respond that if it ordinarily seems to be the case that God has virtues (as [12] suggests), if there are no empirical–logical reasons not to believe this, and if there are no other relevant considerations, then these facts *constitute* a justification for believing that God has virtues. There need not be any further justification or basis for this belief. Indeed, the assumption that there must be some further justification for this belief meets the same fate as the assumption that the sequence (8)–(11) is invalid, namely, that it turns out to be a self-referentially unjustifiable belief by virtue of its implicit reliance on p. The same would hold for the belief that there is any other invalid substitution instance of this argument form.

In summary, I think the translation method can survive the threat posed by the skeptical form of the a priori nonsemantic tenseless theory of time. The tenser can soundly argue that

(16) Ordinary A-sentences and their tokens are untranslatable by B-sentences and their tokens, and this is best explained by the fact that A-sentences ascribe A-properties.

(17) Some A-sentence-tokens are ordinarily accepted as true.

(18) McTaggart's paradox and science do not provide any logical or empirical reasons for rejecting the A-property theory.

(19) There are no other relevant considerations.

Therefore,

(20) It is more reasonable to believe, than not to believe, that the A-property theory is true.

1.5 The A Posteriori Version of the Nonsemantic Tenseless Theory of Time

Suppose that I can establish two things:

(1) Ordinary A-sentences ascribe A-properties to events.

(2) There are no logical difficulties, such as those alleged by McTaggart, entailed by the tensed theory of time.

This would not justify the conclusion that events possess A-properties, since there might be a decisive scientific objection to the tensed theory, one that shows this theory to be, although logically possible and a part of common sense, probably false. There are two sorts of scientific objections that can be made:

1. The tensed theory of time is logically incompatible with a scientific theory that is probably true.

2. The tensed theory of time is not implied by any scientific theory, and it would be implied if it were true.

Objection 1 is usually made in reference to Einstein's theory of relativity. J.J.C. Smart makes the objection that the tensed theory of time cannot be represented on Minkoswki's space–time diagram[28] and Grünbaum makes a similar point in his "Carnap's Views on the Foundation of Geometry":

> The relativistic picture of the world makes no allowance for such a division [of events into "those that have already 'spent their existence', as it were, those which actually exist, and those which are yet to 'come into being'"]. It conceives of events not as "coming into existence" but as simply being and thus allowing us to "come across" them and procure "the formality of their taking place" by our "entering" into their absolute future.[29]

I believe, however, that Howard Stein, Storrs McCall, William Lane Craig, and D. Dieks have plausibly argued that the tensed theory of time is compatible with relativity theory.[30] Further arguments to this effect will be advanced in chapter 7. Accordingly, I shall not respond to objection 1 in this section but only to objection 2. I will respond to objection 2 by showing that the use of the translation method in the next several chapters will be able to undermine one of the crucial premises in the B-theorists's argument for it.

Objection 2 is broader than the objections that have in fact been made by proponents of the B-theory, since it refers to any science, and the objections that have in fact been made refer only to physics in a broad sense (including cosmology). Grünbaum, Russell, Hugo Bergmann, Smart, Quine, David Parks, W. Sellars, H. N. Castañeda and many others believe that physics represent events merely as B-related and therefore that A-properties are not fundamental to physical events. For example, Grünbaum writes:

> It seems to me of decisive significance that no cognizance is taken of nowness (in the sense associated with becoming) in any of the extant theories of physics. If nowness were a fundamental property of physical events themselves, then it would be very strange indeed that it could go unrecognized in all extant physical theories *without detriment to their explanatory success*. And I hold with Reichenbach that "If there is Becoming [independently of awareness] the physicist must know it."[31]

Grünbaum's argument has this structure:

(3) If F is a fundamental property of physical events, F is recognized in some physical theories.

(4) Presentness is unrecognized in all physical theories.

Therefore,

(5) Presentness is not a fundamental property of physical events.

But this argument is unclear at best, since "fundamental" and "recognized" are used but not defined, and definitions that make the argument valid are not easy to come by. But let us waive these difficulties and suppose for the sake of argument that it is true both that (1) presentness is a mind-independent property of physical events only if presentness is recognized in some physical theories and (2) a property F is recognized by a physical theory T only if "F" or a synonym or a definite description of F is used in the statement of T. This supposition is consistent with the views of the proponents of the a posteriori version of the nonsemantic tenseless theory, for they appeal to the principle that

> P Presentness and other A-properties are recognized in physical theories only if A-expressions are used in physical theories

and then argue (or rather, dogmatically state) that A-expressions are not used in physical theories. Thus, Russell avers that "no egocentric particulars [which includes A-expressions] occur in the language of physics."[32] It is alleged that if words like *now* do occur in physical theories, they are used tenselessly to denote an arbitrarily selected B-position. In the words of David Parks, "There is no theory, even the theory of the cosmos as a whole, in which 'now' is any more than a word that might be used to denote any value of *t,* selected at random."[33] Castañeda states the position as follows, using "objective science" and "the science of objective facts" in a way that includes all physics and cosmology: "To say it right away, the great divide in language between the sciences of objective facts and the science of the structure of experience has to do with demonstrative and indexical reference. All indexical reference falls on the side of experience. The realm of objective science has no room for demonstrative or indexical language."[34]

The response to the argument sequence (3)–(5) that I shall give in the following pages aims to do only one thing, namely, to refute the thesis that A-expressions are not used in the statement of physical theories. This refutation does not refute the sequence (3)–(5) but it does refute a part of the B-theorists's justification of (4). The B-theorists justify (4) by appealing to P and the alleged fact that A-expressions are not used in physical theories; consequently, if I can show that A-expressions are used in physical theories, I will nullify a part of their reason for believing (4). But I want to emphasis that showing that A-expressions are used in the physical sciences does *not* entail that A-properties are recognized in the physical sciences, since this usage is consistent with the correctness of the old or new semantic version of the tenseless theory of time, which states that the meaning of A-expressions is identical with, or given by, B-expressions. If the new or old semantic version of the tenseless theory of time is correct, then A-expressions, no less than B-expressions, ascribe, or refer to, only B-relations or B-positions, and consequently, the use of A-expressions in the physical sciences poses no threat to the tenseless theory. *However,* if A-expressions are used to ascribe A-properties, then their use in the physical sciences shows that these properties are recognized in these sciences and the science-based B-theory is thereby impugned. The antecedent of this conditional will be established by my use of the translation method in the following chapters; in this section I wish only to establish that this result of my employment of the translation method will threaten the a posteriori nonsemantic tenseless theory by virtue of the fact that A-expressions *are* used in the physical sciences.

I shall focus only on one branch of physics, the theory of the large-scale structure and evolution of the universe, that is, *cosmology*. Cosmology divides into two branches, theoretical cosmology and observational cosmology. The cosmological theory that dominates current thinking is big bang cosmology.[35] Its theoretical portion consists mainly of Friedmann's solutions to the so-called "Einstein equation" of the general theory of relativity; and its observational portion consists largely of descriptions of or pertaining to the values of the constants or parameters in Friedmann's equations. Friedmann's equations, with the cosmological constant omitted, are

$$-3(d^2a/dt^2) = 4\pi G(p + 3P/c^2)a$$
$$3(da/dt)^2 = 8\pi Gpa^2 - 3kc^2.$$

To determine the values relevant to this equation, we need to know the density p and pressure P of matter, the curvature k of the universe, the scale factor a of the radius of the universe at a given time, the expansion or contraction rate of the spatial surfaces of homogeneity and isotropy (i.e., the value of da/dt), the acceleration or deceleration of the expansion or contraction (i.e., d^2a/dt^2), and Newton's gravitational constant G and the velocity of light c. Now my contention is that although A-expressions are not used in the theoretical portion of big bang cosmology, which consists mainly of the Friedmann equations, they are used in the observational part. The observational portion of big bang cosmology aims to fix the values pertinent to Friedmann's equations, and its problems are explicitly defined relative to the *present* state of the universe. For instance, the cosmologists John Ellis and Keith Olive write:

> The list of cosmological problems to be investigated has often included the homogeneity, isotropy, flatness and age of the *present-day* Universe. Why is the Universe *now* so homogeneous and isotropic on scales which *have only recently* entered our horizon and *were previously* held incommunicado by conventional big-bang cosmology? Why is our Universe nearly flat, having an energy density within 1 or 2 orders of magnitude of the critical closure density, despite *having lived* to the old age of $0(10^{61})$ Planck times? Why is the *present-day* vacuum energy density so small?[36]

In conformity with this approach, Dennis Sciama describes the density p of the universe not tenselessly or solely in terms of B-expressions but in terms of present-tense A-expressions:

> In its *present* state the universe is far too dilute to be able to thermalize radiation in the time available (10^{10} years). Since it is difficult to see how radiation could have been produced already thermalized, we conclude that at sometime in the *past* the universe must have been sufficiently dense to thermalize radiation in the time-scale then prevailing. According to the standard cosmological models this would require a universal density of at least 10^{-14} gm cm^{-3} (that is about 10^{15} times larger than the *present* mean density).[37]

Likewise, P.A.M. Dirac introduces a physical value with the A-expression "The *present* velocity of recession is 10^{-3}";[38] and I. D. Novikov writes of the isotropy and homogeneity of the universe: "Observations primarily of the microwave back-ground radiation show that the Universe expands isotropically with a high degree of accuracy at the *present* time and that the matter distribution is homogeneous on a large scale. . . . This is valid for at least some period in the *past* too."[39]

Examples such as these, which could be multiplied indefinitely, show that David Parks is mistaken in his claim that "now" or "present" is only used in cosmological theory to denote a randomly related time. That Parks is mistaken is shown with especial vividness by R. H. Dickie's use of "the present value of T,"[40] where T is the Hubble age (the age of the universe, how long it has been expanding since the big bang). The present value of T is about fifteen billion years. If some cosmologist used "the present value of T" to refer to some other time, some randomly selected time such as 10^{-43} second after the big bang, she would be accused of misusing scientific terminology, of using the cosmological A-expression "the present value of T" as if it were a B-expression like the temporal variable "t" that appears in the Lorentz transformations of the Special Theory of Relativity.

Suppose, now, that the scientifically minded B-theorist objects that I am here confusing the semantic and pragmatic elements of cosmology. He will allow that A-expressions are used pragmatically in the course of observationally discovering that physical entities have certain properties but deny that these expressions are used semantically in the theoretical statement of these properties. Since only semantic, rather than pragmatic, statements are *parts* of physical theories, he will conclude that my examples have failed to show that A-expressions are parts of physical theories. The B-theorist might accordingly direct against my account the charge that Grünbaum makes in a related discussion, namely, that it

> falsely equates and confuses two *different* meaning components of terms in physics: the physical or semantical with the psychological or pragmatic. The semantical component concerns the properties and relations of purely physical entities which are denoted (named) by terms like velocity. On the other hand, the pragmatic component concerns the activities, both manual and mental, of scientists in *discovering* or coming to *know* the existence of physical entities exhibiting the properties and relations involved in having a certain velocity.[41]

If the B-theorist makes this objection against my account, I respond that he is confusing the distinction between the theoretical and observational components of cosmology with the distinction between the semantic and pragmatic components of cosmology. All of the quoted cosmological statements belong to the semantic com-ponent of observational cosmology. The pragmatic component of observational cosmology is expressed in such sentences as "I am now rotating the telescope to a different angle" and "There is a prominent absorption feature in this spectra-graph right where I expected." The pragmatic statements do not belong to the big bang cosmological theory but merely to the activities involved in verifying this theory. But the quoted expressions such as "The present velocity of recession is 10^{-3}" are *parts* of big bang cosmology, its observational part. They belong to the sentences that state "the properties and relations of purely physical entities which are denoted (named) by terms like velocity"; for "the present velocity of recession is 10^{-3}"

states that velocity of recession presently exemplified by galactic clusters has the value of 10^{-3}, and statements about this velocity are parts of the big bang theory itself. To deny that such expressions belong to the semantical component of cosmology is tantamount to denying that observational cosmology belongs to cosmology, which is absurd.

Let me further quote Grünbaum himself to buttress my point, for Grünbaum refers to a statement in *observational* cosmology as an example of a semantical statement. He notes that "cosmogonic hypotheses make reference to the velocities of masses during a stage in the formation of our solar system which *preceded* the evolution of man and his psychological time sense."[42] If these observational-cosmological hypotheses are semantical, then so is the hypothesis that "the present velocity of recession is 10^{-3}."

If the B-theorist now retreats to the position that A-properties are recognized in the physical sciences only if A-expressions are used in the theoretical portions of these sciences, I respond that her position is preposterous, amounting to the claim that a certain property the mention of which properly belongs to observational physics is recognized by physics only if it is mentioned in theoretical physics. Suppose I held that the specific location of the Local Group vis-à-vis the other galactic clusters in the universe "is not recognized in cosmology" since it is not mentioned in theoretical cosmology! The spatial location of the Local Group in the universe, as well as the temporal location of presentness (i.e., which physical state has the A-position of presentness) *by definition* are observational matters.

The following claim is not an objection to the position for which I have been arguing in this section.

(6) The A-expressions in observational cosmology are one and all superfluous, since they can be replaced *salva sensu* by B-expressions.

Item (6) is consistent with the position that *A-expressions belong to observational cosmology,* which is all that I have been concerned to argue. But it is inconsistent with the position for which I am going to argue in the next three chapters, namely, that A-expressions cannot be replaced *salva sensu* by B-expressions and ascribe A-properties.

Before I begin these arguments, I need to make some remarks about the alleged relativity of time. If A-properties are possessed relatively to a reference frame, these properties are not monadic but dyadic properties of such forms as ( )'s being present relatively to ( ) and ( )'s being past relatively to ( ), where the second parentheses are filled in by expressions denoting a reference frame. However, I shall argue in chapter 7 that time is absolute and that Einstein's theory implies merely that observable luminal relations among physical events are possessed relatively to a reference frame. I will argue that

(7) If light signals were sent from events E_1 and E_2, they would be observed (by a hypothetical observer) to arrive simultaneously at the observationally verifiable midpoint between E_1 and E_2

neither means, nor entails, nor is entailed by

(8) E_1 and E_2 are simultaneous

and therefore that if (7) is true only relatively to a reference frame that does not entail (8) is true only relatively to a reference frame. I shall also provide a positive argument that simultaneity and other temporal relations are absolute, even though in cases of distant physical events we cannot find out which absolute temporal relations hold. Accordingly, my discussions of time in chapters 2–6 should be understood in terms of the thesis for which I shall argue in chapter 7, namely that B-relations are dyadic properties (rather than triadic properties whose third term is a reference frame) and A-properties are monadic properties (rather than dyadic properties whose second term is a reference frame).

Notes

1. References are given in note 12.

2. My use of "event" is thus broader than Jaegwon Kim's ("Events as Property Exemplifications," in *Action Theory,* ed. M. Brand and D. Walton [Dordrecht Reidel, 1976]) and Donald Davidson's (*Essays on Actions and Events* [Oxford: Clarendon, 1980]). As far as I know, only Nicholas Wolterstorff has used "event" in the broad sense I am adopting. See his "God Everlasting," in *God and the Good,* ed. C. Orlebeke and I. Smedes (Grand Rapids, MI: Eerdmans, 1975), 181–203, and "Can Ontology Do Without Events?," in *Essays on the Philosophy of Roderick M. Chisholm,* ed. Ernest Sosa (Amsterdam: Rodopi, 1979), 177–204. It should be emphasized that this technical use of "event" is much broader than the ordinary use of "event" and makes no claim to capture ordinary usage.

3. Condition iv entails that I do not count as A-sentences the five traditionally undiscussed but conventionally standard uses of tensed sentences that I described in "The Multiple Uses of Indexicals," *Synthese* 78(1989): 167–191. A discussion of these uses would unnecessarily complicate the theory I am developing.

4. References to these philosophers' writings are given in chapter 5, note 12, where I critically discuss their versions of the tensed theory of time.

5. See my "Problems with the New Tenseless Theory of Time," *Philosophical Studies* 52(1987): 77–98, and "The Co-reporting Theory of Tensed and Tenseless Sentences," *Philosophical Quarterly* 40(1990): 223–32.

6. Alfred J. Ayer, *The Problem of Knowledge* (Harmondsworth: Penguin Books, 1956); Milton Fisk, "A Pragmatic Account of Tenses," *American Philosophical Quarterly* 8(1971): 93–98; Nelson Goodman, *The Structure of Appearance* (Cambridge: Harvard University Press, 1951); W.V.O. Quine, *Word and Object* (Cambridge: Massachusetts Institute of Technology Press, 1960); J.J.C. Smart, *Philosophy and Scientific Realism* (London: Routledge & Kegan Paul, 1963); Clifford Williams, "'Now', Extensional Interchangeability, and the Passage of Time," *Philosophical Forum* 5(1974): 405–23.

7. Richard Gale, *The Language of Time* (New York: Humanities, 1968), 38.

8. Ibid., 55.

9. Ibid., 38–39.

10. It should be added that after presenting the argument just discussed, Gale proceeds to "drive home" its point by considering some examples of how A-sentences are used in daily

life (ibid., 56ff.); but in his discussion of these examples, Gale retains his question-begging assumption that A-sentences express statements that refer to events with A-determinations. Nicholas Wolterstorff reproduces and endorses an abbreviated version of Gale's informally invalid argument in his "Can Ontology Do Without Events?" George Schlesinger's approach to this argument is more sensible, because he criticizes the argument in his *Aspects of Time* (Indianapolis: Hackett, 1980), 28–30. Unfortunately, however, Schlesinger's implicit point (in this and other discussions) seems to be the erroneous contention that there are no linguistic matters of fact that could decide the issue of whether or not A-sentences are translatable and hence that a "semantic ascent" is of no avail. I criticize Schlesinger's views on this issue in chapter 3.

11. See my "Problems with the New Tenseless Theory" and "The Co-reporting Theory."

12. See Stephen Braude, "Tensed Sentences and Free Repeatabity", *The Philosophical Review* 82 (1973): 188–214; J.J.C. Smart, "Time and Becoming," in *Time and Cause*, ed. Peter van Inwagen (Boston: Reidel, 1980), 3–15; D. H. Mellor, *Real Time* (Cambridge: Cambridge University Press, 1981); Murray MacBeath, "Mellor's Emeritus Headache," *Ratio* 25(1983): 81–88; Paul Fitzgerald, Review of *Real Time* by D. H. Mellor, *Philosophy and Phenomenological Research* 45(1984): 281–86; Michelle Beer, "Temporal Indexicals and the Passage of Time," *Philosophical Quarterly* 38(1988): 158–64; Michelle Beer and Richard Gale, "An Identity Theory of the A- and B-Series," *Dialectics and Humanism* (forthcoming).

13. Smart, "Time and Becoming," 5.

14. Fitzgerald, Review of *Real Time* by D. H. Mellor, 281.

15. See L. Nathan Oaklander, "The New Tenseless Theory of Time: A Reply to Smith," *Philosophical Studies* 58(1990): 287–292. This is a reply to my "Problems with the New Tenseless Theory."

16. Oaklander, "New Tenseless Theory of Time," 290.

17. See my "Problems with the New Tenseless Theory."

18. Smart, "Time and Becoming," 11.

19. Ibid.

20. Ibid., 15.

21. Mellor, *Real Time*, 74.

22. Ibid., 75.

23. Ibid., 76.

24. Ibid., 87.

25. Oaklander, "New Tenseless Theory of Time," 291.

26. Idem, *Temporal Relations and Temporal Becoming* (New York: 1984), 130.

27. For a further explanation and justification of the contradictory of this skeptical principle, see my *Felt Meanings of the World: A Metaphysics of Feeling* (West Lafayette, Ind.: Purdue University Press, 1986), 131–34. Panayot Butchvarov has claimed that the nonskeptical principle that I advocate is unsound or unreliable, but he makes no attempt to defend his claim against my argument that claims of his sort are implicitly self-contradictory. See Panayot Butchvarov, Review *The Felt Meanings of the World* by Quentin Smith, *Nous* 23(1989): 281–84, esp. p. 283.

28. J.J.C. Smart, to author, December 1986.

29. Adolf Grünbaum, "Carnap's Views on the Foundation of Geometry," in *The Philosophy of Rudolp Carnap* ed. Paul Schilpp (Lasalle: Open Court, 1963), 659.

30. See Storrs McCall, "Objective Time Flow," *Philosophy of Science* 43(1976): 337–62; William Lane Craig, "God and Real Time," *Religious Studies* 26(1990): 335–47;D. Dieks, "Special Relativity and the Flow of Time," *Philosophy of Science* 55(1988): 456–60; Howard Stein, "On Einstein-Minkowski Space-Time," *The Journal of Philosophy* 65 (1968): 5–23.

31. Adolf Grünbaum, *Modern Science and Zeno's Paradoxes* (Middletown, CT: Wesleyan University Press, 1967), 20.

32. Bertrand Russell, *An Enquiry into Meaning and Truth* (London: Allen & Unwin, 1940), 108.

33. David Parks, "The Myth of the Passage of Time," in *The Study of Time,* ed. J. T. Fraser et al. (New York: Springer–Verlag, 1972), 112.

34. Hector-Neri Castañeda, "Reference, Reality, and Perceptual Fields," in *Proceedings of the American Philosophical Association* 53(August 1980): 769–70.

35. For a presentation and philosophical discussion of big bang cosmology, see Quentin Smith, "Anthropic Principle and Many-Worlds Cosmologies," *Australasian Journal of Philosophy* 63(1985): 336–48; idem, "World Ensemble Explanations," *Pacific Philosophical Quarterly* 67(1986): 73–86; idem, "The Uncaused Beginning of the Universe," *Philosophy of Science* 55(1988): 39–57; idem, "A Natural Explanation of the Existence and Laws of Our Universe," *Australasian Journal of Philosophy* 68(1990): 22–43; idem, "Atheism, Theism, and Big Bang Cosmology," *Australasian Journal of Philosophy* 69(1991): 48–66; and idem, "A Big Bang Cosmological Argument for God's Nonexistence," *Faith and Philosophy* 9(1992): 217–237.

36. John Ellis and Keith Olive, "Inflation Can Solve the Rotation Problem," *Nature,* June 23, 1983, p. 679. Italics in notes 36–39 are mine.

37. Dennis Sciama, "The Universe as a Whole," in *The Physicist's Conception of Nature,* ed. Jagdish Mehra (Dordrecht: Reidel, 1973), 24–25.

38. P.A.M. Dirac, "Fundamental Constants and their Development in Time," in *The Physicist's Conception of Nature,* ed. Jagdish Mehra (Dordrecht: Reidel), 47.

39. I. D. Novikov, "Isotropization of Homogeneous Cosmological Models," in *Confrontation of Cosmological Theories with Observational Data,* ed. M. S. Longair (Dordrecht: Reidel, 1974), 273.

40. R. H. Dickie, "Dirac's Cosmology and Mach's Principle," *Nature,* November 4, 1961, p. 440.

41. Adolf Grünbaum, "Carnap's Views on the Foundation of Geometry," in *The Philosophy of Rudolp Carnap,* ed. P. Schilpp (Lasalle: Open Court Press, 1963), 630.

42. Ibid.

2

The Untranslatability of A-Sentences by Tenseless Date-Sentences

2.1 The Date Theory of A-Sentences

There are two main types of B-sentences, *date-sentences* and *token-reflexive sentences*. An example of the former is "John (is) born earlier than May 1, 1960," and of the latter, "John (is) born earlier than this utterance." The distinction between these two types provides a means for classifying most of the proponents of the old or new versions of the semantic tenseless theory of time. Some of these proponents hold the *date-theory* of A-sentences, that tokens of A-sentences are translated (in some sense) by tenseless date-sentences, or at least have truth conditions that are completely statable by these date-sentences. Other proponents hold the *token-reflexive* theory of A-sentences, that these sentences are translated (in some sense) by token-reflexive tenseless sentences, or at least have truth conditions that are completely statable by these token-reflexive sentences. In this chapter, I shall argue against the date-theory of A-sentences, in chapter 3, against the token-reflexive theory; in chapter 4, I shall argue against the other less frequently espoused versions of the semantic tenseless theory and develop a positive theory of A-sentences that conforms to the tensed theory of time. This will prepare the way for my defense of the presentist theory that every sentence, including every tenseless sentence, ascribes a presentness-involving property (see part II).

The following quotations convey the flavor of the date-theory of A-sentences:

"Fido is [was, will be] running" as said at a given time, say t_0, expresses the very same proposition expressed by "Fido (is) running at [before, after] t_0." (Milton Fisk)

Take any token of the form

(i) Event E is occurring now.

Translate it, or if you prefer, the proposition which it expresses, by

(ii) Event E *occurs* (tenseless) at (time) theta.

27

Here the proper name "theta" is what we might call a "non-descriptively individuat-
ing proper name" of the time in question, that is, the time at which the token to be
translated was produced. (Paul Fitzgerald)

A certain "ran" is translated by any "runs (tenseless) on Jan. 7, 1948 at noon
E.S.T." (Nelson Goodman)

When I say

(4) I was insulted yesterday,

a specific content—*what I said*—is expressed. Your utterance of the same sentence, or mine on another day, would not express the same content. It is important
to note that not just the truth value may change; what is said is itself different.
Speaking today, my utterance of (4) will have a content roughly equivalent to that
which

(5) David Kaplan is insulted on April 20, 1973

would have when spoken by you or anyone at anytime. (David Kaplan)

Only at twelve noon can someone think the thought consisting of noon and the
sense of "The meeting starts at ( )" by entertaining the sense of "the meeting starts
now." (John Perry)

"Henry Jones of Lee St., Tulsa, is ill" . . . uttered as a tensed sentence on July 28,
1940, corresponds to the *statement* "Henry Jones of Lee St., Tulsa, is [tenseless] ill
on July 28, 1940." (W.V.O. Quine)

When we are told "Mrs. Brown is not at home," we know the time at which this is
said, and therefore we know what is meant. But in order to express explicitly the
whole of what is meant, it is necessary to add the date, and then the statement is no
longer "variable," but always true or always false. (Bertrand Russell)

The proposition expressed at any given time, t_n, through the use of a sentence in
which "now" is used is the same as the proposition expressed at t_n through the use
of a sentence formed by replacing the "now" in the first sentence with any date-
expression that is used to refer to t_n. (Clifford Williams)[1]

Other philosophers who subscribe to some version of the date-sentence theory
include Joseph Almog, Michelle Beer, Tyler Burge, Cohen, Nagel, Lynne Rudder
Baker, Frege, William Lycan, Jerrold Katz, Nelson Pike, Mark Richards, Jan David
Wald.

The date theory of A-sentences is either an *indirect reference* theory or a *direct
reference* theory, I will explain the versions of these theories that postulate *de dicto*
and *de re* date-propositions, respectively, as the semantic content of A-sentence-
tokens and will at the end of these explanations say something about the nominalist
versions of the date theory, such as Quine's.

The indirect reference theory holds that A-words (the tensed copulae and verbs
and the temporal adverbs, adjectives, and pronouns in A-sentences) refer to times
indirectly via definite descriptions of them. These definite descriptions are not, of
course, "definite descriptions" in the sense of linguistic phrases, since these phrases
are not parts of the relevant A-sentences; rather, they are "definite descriptions" in
the sense of components of the propositions expressed by the A-sentences. But
strictly speaking, the A-words express these propositional definite descriptions not
by themselves but only with the help of their contextual features. Frege makes this

point about A-sentences and other sentences containing indexicals: "The mere word-ing, as it is given in writing, is not the complete expression of the thought, but the knowledge of certain accompanying conditions of utterance, which are used as a means of expressing the thought are needed for its correct apprehension."[2] The "time of utterance," Frege adds, is the relevant condition that helps to express the descriptive sense of the A-words: "If a time indication is needed by the present tense one must know when the sentence was uttered to apprehend the thought correctly. Therefore the time of utterance is part of the expression of the thought."[3]

In order to make this theory more precise, it will be helpful to provide defini-tions of times and dates and delimit the nature of the corresponding propositional descriptions. First of all, let us give the two standard definitions of times, the definitions used in the *reductionist* and *substantival* theories. The reductionist theo-ry is often called the "relational" theory, and the substantival theory, the "absolutist" theory; but since "relational versus absolutist theories of time" can mean many different things (John Mackie distinguishes fourteen different senses),[4] it is best to use different terminology to avoid ambiguity. The reductionist theory of time defines a time as a *set of simultaneous events*. W. H. Newton-Smith defines the "reduction-ist" theory's conception of a moment: "We can form the collection of all events simultaneous with any particular event used in identifying the moment. This collec-tion, the reductionist claims, just is the moment. To say that such and such an event occurred at such and such a moment is just to say that the event in question is a member of the set of events constituting the moment."[5] Russell espouses a reduc-tionist theory of time in his later writings and defines "instant" in a related way: "An 'instant', as I propose to define the term, is a class of events having the following two properties: (1) all the events in the class overlap; (2) no event outside the class overlaps with every member of the class."[6] Examples of reductionist theories of time beside Russell's are Hans Reichenbach's, Adolf Grünbaum's, Mario Bunge's and Ian Hinckfuss's.[7]

The substantival theory, on the other hand, defines a time as a particular item that is logically independent of the events (if any) that stand to it in the relation of *occupying* it. Let us depart from Newton-Smith's terminological usage and reserve the word *moment* for a time in the substantival sense. For any given moment M, if M is occupied by some simultaneous events, there is a set S of all these occupying events that is both distinct from M and accidentally related to M. If we suppose that in the actual world, M is occupied by the members of S, there will be some merely possible world in which M is occupied by the members of some other set S* of simultaneous events, and there will be some still other world (containing "empty time") in which M is not occupied by any events at all (or at least not by any "events" in the narrow sense that involves physical or mental changes or states). Newton-Smith calls these substantival moments "temporal items" and says that (what I am calling) the substantivalist thesis is that "the existence of temporal items is ontologically independent of the existence of things in time. Temporal relations between things in time hold in virtue of temporal relations holding between the times at which the things in time occur."[8] Examples of substantivalist theories are Isaac Newton's, Richard Swinburne's, Sydney Shoemaker's, and the theory I de-fended in *The Felt Meanings of the World: A Metaphysics of Feeling*.[9]

There are some nonstandard definitions of times, such as A. N. Prior's definition of a time as a conjunction of true propositions[10]; but these nonstandard definitions will make no material difference to my argument, and I shall not discuss them further.

If a time is either a set of simultaneous events or a moment, then a *date* is definable as a set of simultaneous events (or moment) that stands in a B-relation to the set of simultaneous events (or moment) at which Christ's birth is (conventionally said to be) located. (Obviously, the zero-point of a dating system can be located at any event; but I am here defining "date" in terms of the system we conventionally use.) On the reductionist theory, an example of a date is the set S_1 of simultaneous events that bears to the set S_0 of simultaneous events that includes Christ's birth the relation of *being 1,985 years, 56 days, 10 hours, and 3 minutes E.S.T. later than it.* Some philosophers think it nonsensical (or at least false) to talk about sets as sustaining temporal relations ("sets are abstract objects, and abstract objects by definition are timeless"); but the idea that sets of simultaneous events do not exist simultaneously with their members seems counterintuitive, and I agree with David Lewis that "a set of [spatiotemporally] located things *does* seem to have a [spatiotemporal] location: it *is* where [and when] its members *are*."[11] But this is not a mere matter of stipulative definition or intuitive seemings; I will argue in chapter 6 that each object, concrete or abstract, possesses temporal relations and properties.

The substantival theory of times represents dates as moments that exemplify date-properties; for example, the moment M_1 is a date, since it has the property of *being 1,985 years, 56 days, 10 hours, and 3 minutes E.S.T. later than the moment M_o, which is occupied by Christ's birth.* For brevity's sake, I will sometimes talk of a date as a set or moment that bears such a B-relation to Christ's birth.

Fortified with these definitions of times and dates, we can give further precision to the indirect reference date-theory of A-sentences. The theory is that A-expressions express propositional definite description of dates not by themselves but in conjunction with the date at which the A-expressions are tokened. Each A-expression-token has the property of the form *occurring at D,* where D stands for a date. It is this property that is the accompanying condition that (along with the A-expression token) *expresses* the propositional definite description of D. The relevant A-expression-tokens also introduce additional senses into the proposition expressed; for example, a token of "yesterday" introduces, in addition to the propositional description of the date D at which it occurs, the sense *whatever day is one day earlier than (D).* Thus, if I utter "Yesterday was cloudy" this utterance has the property, say, of *occurring at the daylong moment (or set of events) that is 1,988 years and 90 days later than Christ's birth;* and this property, in conjunction with the utterance of "Yesterday," expresses the sense *whatever daylong moment or (set of events) is one day earlier than the daylong moment (or set of events) that is 1,988 years and 90 days later than Christ's birth.*

Some defenders of the new tenseless theory of time might wish to say that A-sentence-tokens and their date-properties do not express such senses but that B-sentences expressing them suffice to give the truth conditions of the A-sentence-tokens. I shall show that this is not the case in the course of my argument that these senses are not expressed by A-tokens (see section 2.2).

But before I begin these arguments, I want to explain the direct reference date theory of A-sentences, which states that A-expression-tokens do not express propositional definite descriptions of dates but refer directly to the dates. These tokens introduce the dates themselves, rather than a descriptive sense into the proposition expressed by the A-sentence-token. The *de re* proposition expressed by a noon occurrence of "The meeting starts now" includes the meeting itself (supposing that the incomplete definite description "The meeting" is used referentially, rather than attributively), the property of starting and the set of events or moment that is identical with noon. This proposition contains all and only the members of the set

(S) [The meeting, the property of starting, noon]

where "The meeting" and "noon" are used as directly referring terms.

But as this example arguably suggests, the direct reference date-theory is too rigid as it stands to be even prima facie plausible, since the claim that A-expression-tokens introduce no senses at all into the proposition but merely the date is overtly inconsistent with the syntactic features of certain A-locutions. I am not here making a criticism of the direct reference theory from the point of view of the tensed theory of time (as I shall later) but an internal criticism that presupposes only the tenseless theory. Consider the difference between a pronomial and adverbial use of "now," the first being illustrated by "Now is when the meeting starts," the second, by "The meeting starts now." The direct reference date theory runs into no internal problems with the former sentence, which it translates as "Noon is when the meeting starts" for a noon utterance of this sentence, with "noon" referring directly to noon. But incoherence appears if this interpretation is applied to a noon utterance of "The meeting starts now," for this would translate as the syntactically incomplete sentence "The meeting starts noon." A syntactically complete version of this sentence would read "The meeting starts *at* noon"; but the translation of "now" by "at noon" is forbidden by the direct reference principle that "now" *merely* refers directly to times and does not have any predicative sense, such as the relational sense *at* (*simultaneously with*).

It is instructive to consider how some of the direct reference date-theorists fudge or gloss over this issue. John Perry gives this account of the aforementioned sentence where "now" is used adverbially: "Only at twelve noon can someone think the thought [proposition] consisting of noon and the sense of 'The meeting starts at ( )' by entertaining the sense of 'the meeting starts now'."[12] Notice that Perry introduces the sense expressed by "at" into the *analysans* of the proposition expressed at noon by "the meeting starts now," even though his next premise forbids its introduction: "The demonstrative ['now'] in context gives us the [object, noon itself], the rest of the sentence the [incomplete sense]."[13] If this premise is true, then the demonstrative cannot contribute a part of the incomplete sense (that expressed by "at") and the *analysans* should consist only of the rest of the sentence ("The meeting starts") and a word that merely directly refers to the object ("noon").

A defender of the direct reference date theory might respond that the sense of "at" is contributed by the verb "starts," which means "starts at." But this is plainly false, since if "starts" means "starts at," then these two expressions are intersubstitutable *salvu sensu,* which they are not. "Now is when the meeting starts" is so far from being synonymous with "Now is when the meeting starts at" that the former alone is syntactically well formed. The "direct reference" theorist might then respond that verbs like "start" are systematically ambiguous, having different senses when they are used with demonstrative adverbs than when not so used. When they are so used, they acquire an additional sense that relates the rest of their sense to the direct referent of the demonstrative, so that "starts" now expresses what is expressed by "starts at." But this response is implausible and ad hoc. "Starts" is used in the same way, as a verb, in "The meeting starts now" and "Now is when the meeting starts"; and the relevant difference is in the change from a pronominal to an adverbial use of "now." Since the syntactic change involves "now," not "starts," it is intuitively more credible to suppose that the semantic change involves "now" than to suppose it involves the verb. Indeed, this change is precisely what we would expect in a change from a pronominal to an adverbial use; for a pronoun is merely a referring device, and an adverb is a *qualifying* expression, something that characterizes what is expressed by a verb. As a qualifying expression, the adverb expresses a property of the semantic content of the verb. It is natural to think that the adverbial use of "now" expresses the relational property *being simultaneous with noon,* which includes noon itself as a constituent, and that the function of this adverb is to convey that this property is possessed by the meeting's start. We may still allow, consistently with the direct reference theory, that this adverb refers to noon directly, rather than through a propositional definite description; but we deny that this reference is its sole semantic property, since the adverb, in addition, ascribes to the meeting's start a property partly composed of its direct referent, *being simultaneous with noon.*

This internal modification of the direct reference date theory seems to me to be desirable, and I will assume it in what follows.[14] It prejudges nothing against the tenseless theory of time; but it will help prepare the way for my introduction (in chapter 4) of additional senses into A-words like "now," namely, A-senses.

I have explained the date theory of A-sentences in terms of the propositionalist theories; but the nominalist version deserve a word, as well. According to nominalist date theorists, such as Quine, we do not need to postulate *de dicto* or *de re* propositions as semantic contents of the A-sentence-tokens but can explain this content solely in terms of the eternal B-sentences that correspond to these tokens:

> But instead of appealing here to propositions, or meanings of eternal sentences, there is no evident reason not to appeal simply to the eternal sentences themselves as truth vehicles. If we undertake to specify the proposition "expressed" by the utterance of some noneternal sentence, e.g., "The door is open," on some particular occasion, we do so by bracketing some eternal sentence that means the proposition; thus we have had to compose an appropriate eternal sentence anyway, and we could as well stop there.[15]

The eternal sentences corresponding to A-sentence-tokens include linguistic definite descriptions that are "explicit references to dates or periods."[16] These references are "left tacit"[17] in the use of A-sentences. This nominalist theory is mistaken if it can be shown that uses of such sentences as "The door is open" do not tacitly refer to dates, which I shall do in section 2.5, using Quine's own liberal criteria for "translation" (or "paraphrase," as he prefers to call it).

The rest of the present chapter consists of five sections. In the four sections that immediately follow, I argue that tenseless date-sentences, however conceived, meet none of my four conditions of translating uses of A-sentences. I begin in section 2.2 by arguing that tenseless date-sentences and A-sentence-tokens have different truth conditions and proceed to argue in section 2.3 that they have different confirmation conditions, in section 2.4 that they are not logically equivalent, and in section 3.5 that they are not logically identical. In section 2.5 I shall concede that they are intersubstitutable *salva veritate* in extensional contexts but note that this is insufficient for them to express the same proposition. I do not believe that any single one of my arguments is sufficient to refute the tenseless date-sentence theory (there is no single knockdown argument). But the cumulative effect of the many arguments I shall present suffices, in my judgment, to refute the tenseless theory. This is the reason that I attack the tenseless theory at many different points, rather than concentrate on a single problem. If my tensed theory is to be completely refuted and the tenseless theory to be completely justified, then each of the many arguments offered in the following sections must be shown to be unsound.

2.2 The Truth Conditions of A-Sentence-Tokens and Tenseless Date-Sentences

Previous attempts to refute the date-sentence theory of A-sentences have concentrated on the date theory developed by proponents of the old tenseless theory of time and have typically consisted of arguments to the effect that A-sentence-types have semantic rules different from those of tenseless date-sentence-types and therefore are untranslatable by them. This is the line of argument developed by Richard Taylor and Richard Gale in his tenser period.[18] However, the arguments of Taylor and Gale have been successfully rebuted by Clifford E. Williams,[19] who pointed out that arguments of this sort miss the point of the date-sentence theory, which does not claim that the date-sentence-type translates (in some sense) an A-sentence-type but that it translates some token of the A-sentence-type. The thesis of the date theory is that each successive token of some A-sentence-type corresponds to a distinct date-sentence-type, such that corresponding to the token of "Henry is ill" that occurs on July 28, 1940 is the date-sentence-type "Henry (is) ill on July 28, 1940" and corresponding to the token of this A-sentence that occurs on July 29 there is "Henry (is) ill on July 29, 1940." Thus, if the date theory of A-sentences is to be refuted, it must be shown that the relevant semantic features of a given token of an A-sentence-type are different from those of the corresponding date-sentence-type. In this section, I shall show that the truth conditions of a given A-sentence-token are different from those of its corresponding date-sentence-type and that this refutes both the old and new versions of the tenseless date-sentence theory.

Let me begin by pointing out that a token of an A-sentence has *token-reflexive* truth conditions, whereas no tenseless date-sentence-type and no token of this type has such conditions (except in exceptional cases to be noted). The truth value of an

A-sentence-token is dependent upon the B-relation of the token to the event the token is about. A token of a future tensed A-sentence is true if and only if it occurs earlier than the event it is about; a token of a present tensed A-sentence is true if and only if it occurs simultaneously with the event it is about; and a token of a past tensed A-sentence is true if and only if it occurs later than the event it is about. For instance, a token of "Henry was ill" is true if and only if it occurs later than Henry's illness. By contrast, when a tenseless date-sentence is used is irrelevant to its truth value. For example, it is false that the date-sentence "Henry (is) ill before April 5, 1989," or any given token of this sentence, is true if and only if it is used later than the event it is about. It is rather the case that the tenseless date-sentence or any one of its tokens is true if and only if the event it is about has the appropriate B-relation to the date the sentence refers to. "Henry (is) ill before April 5, 1989" is true if and only if Henry's illness is earlier than April 5, 1989 and if this condition obtains, the sentence is true regardless of whether it is used before, after, or simultaneously with Henry's illness. Of course, some tenseless date-sentences have token-reflexive truth conditions ("This sentence-token occurs later than Henry's illness and on April 5, 1989" is true if and only if the token occurs later than Henry's illness and on April 5, 1989); but these are abnormal cases and are not the sorts of translating or truth-condition-stating date-sentences the detensers have in mind.

It is important to make explicitly clear that in this contrast I am specifying only *tenseless* truth conditions of the *tenseless* truth values of A-sentence-tokens. It is my contention, for example, that

(1) A token of "Henry was ill" (is) true if and only if this token (is) later than Henry's illness.

By describing the truth conditions of A-sentence-tokens in a tenseless language and by defining them in terms of the obtaining of certain B-relations between the tokens and the events they are about, I am not begging the question against the date-theorist by assuming from the outset that A-sentence-tokens have truth conditions and truth values that involve A-properties and are statable only in a tensed language. I believe that A-sentence-tokens do have such truth conditions and values, for example, that a token of "Henry was ill" is presently true if and only if Henry's illness is now past; but my strategy is not to assume this but to prove it by arguing from the differences in tenseless truth conditions and values (and other tenseless semantic properties, discussed in later sections), between A-sentence-tokens and date-sentences. These tenseless differences are my concern in this chapter.

At first glance, this account of A-sentence-tokens and tenseless date-sentences might seem to show that these locutions have different truth conditions and therefore that date-sentences cannot translate A-sentence-tokens. It might also seem to show that the date theory espoused by proponents of the new tenseless theory of time is mistaken, since if A-sentence-tokens have token-reflexive truth conditions and not date-involving truth conditions, then date-sentences cannot give the truth conditions of A-sentence-tokens. But a refutation of the old and new tenseless date theories is not as easy as this, since some of the proponents of these theories have a seemingly powerful argument that A-sentence-tokens do have date-involving truth conditions—in fact, the very same truth conditions as their corresponding tenseless date-sentences. J.J.C. Smart, holds, for example, that "when a person P utters at a time t the sentence 'Event E is present' his assertion [utterance] is true if and only if E is at t. More trivially, when P says at t 'time t is now' his assertion is true if and only if t is at t so that if P says at t 't is now' his assertion is thereby true."[20]

If the expression "t" is replaced by a particular date-description and "P" by a person's name, we can obtain truth conditions of particular utterances of A-sentences. Using the example of "Henry is ill," we may say

(2) "Henry is ill" is true as spoken by John on July 28, 1940 if and only if Henry (is) ill on July 28, 1940.

The truth condition clause following the biconditional is exactly the same as the clause following the biconditional in the statement of the truth conditions of the corresponding tenseless date-sentence:

(3) "Henry (is) ill on July 28, 1940" is true if and only if Henry (is) ill on July 28, 1940.

Proponents of the old tenseless theory of time will take these claims as evidence that tenseless date-sentences *translate* the A-sentence-tokens, and defenders of the new tenseless theory (such as Smart) will take them as showing that the date-sentences *suffice to give the truth conditions and thereby the meaning* of the A-sentence-tokens, from which it follows that these tokens convey no information about time not conveyed by the date-sentences.

I believe, contrary to these allegations, that A-sentence-tokens do *not* have these date-involving truth conditions. If we call the utterance mentioned in (2) the utterance U, then we may say that (2) states that a necessary condition of U's truth is that Henry be ill on July 28, 1940. In possible-worlds terminology, this means that U is not true in any world in which Henry is not ill on July 28, 1940. But this is not the case. In the actual world, U occurs at a time that possesses the date-property of being 1,939 years, 6 months, and 27 days later than Christ's birth. Let us first suppose that the reductionist theory of times is true and that this time is a set of simultaneous events, two of which are U and Henry's illness. This set of events does not possess the aforementioned date-property in each possible world in which it exists. In one of these worlds, W_1, Christ was not born at all; and in another world, W_2, Christ was born 1,938 years earlier than the set of events that contains Henry's illness. In both of these worlds, U is true, since U is simultaneous with Henry's illness. But "Henry (is) ill on July 28, 1940" is false in W_1 and W_2, since, in these worlds, Henry is not ill at whatever set of events has the property of being 1,939 years, 6 months, and 27 days later than Christ's birth.

The same result follows if we assume the substantival theory of times. In the actual world, U and Henry's illness occupy the moment that has the date-property of being 1,939 years, 6 months, and 27 days later than Christ's birth. But in W_1 and W_2, the moment occupied by U and Henry's illness does not possess this date-property; consequently, the date-sentence—but not U—is false in these worlds.

It is conceivable that a defender of the direct reference date theory would object that my account presupposes an indirect reference theory of A-sentence-tokens and tenseless date-sentences. My remarks presuppose the view that the reference to the date is mediated by a propositional definite description such as that expressed by an attributive use of "the time that is 1,939 years, 6 months, and 27 days later than Christ's birth." Since this definite description refers to whatever time possesses this property, it refers to different times in different worlds, since in different worlds different times possess this property. In the actual world, set S or the moment M possess it; but in some other world, the set S* or moment M* possesses this property and thereby the reference of the description shifts to S* or M*. But suppose the direct reference theory is true, such that the date-sentence stating the truth conditions of U, "Henry (is) ill on July 28, 1940," involves a *referential,* rather than *attributive,* use of "July 28, 1940." In this case, this description directly refers to the set or moment that is, in fact, 1,939 years, 6 months, and 27 days later than Christ's birth and refers to this same set or moment in each possible world in which it exists, regardless of whether or not it possesses this date-property there. If we use "S" ("M") as a directly referential name of this set (moment), we may say that the truth conditions of U are that it is true if and only if Henry (is) ill at S (M).

This direct reference theory provides an adequate response to my scenario involving W_1 and W_2 that assumed the reductionist theory of time, for in each of the worlds in which the set S that actually contains U exists, U is true regardless of whether or not S possesses its actual date-property; for the set S that actually contains U also actually contains Henry's illness and therefore, since sets contain their members essentially, *essentially* contains both events, rendering U true in each world in which this set exists. Consequently, "Henry (is) ill at S," like U, is true in W_1 and W_2.

The direct reference theory's success is limited, however, since there are worlds in which U is true but the referentially used date-description is not. There is a world W_3 in which S does not exist but in which U occupies a set S_1 that contains all and only the events that S contains except that S_1 contains a certain dustlike particle on the planet Venus hitting the ground, whereas S does not. In this world, U is true, since it is simultaneous with Henry's illness, but "Henry (is) ill at S" is false, since Henry is ill at S_1, instead.

If we adopt the substantival theory, the direct reference theory will also fail to give necessary conditions for the truth of U. There is a world W_4 in which U and Henry's illness and all the other events in S (or S_1) occupy M*, rather than M; consequently, U is true, but "Henry (is) ill at M" is false.

I conclude that it is false that U is true if and only if Henry (is) ill on July 28, 1940, regardless of whether this date-description be read attributively or referentially. Tenseless date-sentences do not state the truth conditions of U and other A-sentence-tokens; consequently, the old and new versions of the tenseless date theory are false.

But the tenseless date-theorists are not without response. Smart has responded to an earlier and abbreviated version of my criticism by changing his date theory so that it gives world-indexed truth conditions, rather than truth conditions simpliciter.[21] Smart now holds that it is not the case that "E is present" as spoken by P at t is true if and only if E is at t but that

(4) "E is present" as spoken by P at t in world W is true in W if and only if E is at t in W.

I agree with Smart that (4) is true, but this fails to save the date-sentence theory, since the fact that the non-world-indexed version of (4) is false suffices to refute the theory that date-sentences give the meaning of A-sentence-tokens. World-indexed truth condition sentences do not give the semantic content of utterances, since the criterion of truth for these sentences is that the clause on the right side of the biconditional that has the world-indexed phrase appended to it has the same truth value in the mentioned world as the utterance denoted by the clause on the left side of the biconditional. This allows *any clause* with the same truth value in W as "E is at t" to be substituted for "E is at t" and this prevents (4) from giving the meaning of the utterance. Consider a concrete example,

(5) "Henry is ill" as spoken by John on July 28, 1940 in W is true in W if and only if Henry (is) ill on July 28, 1940 in W.

Suppose that W is the actual world and that the utterance mentioned is true. Then, we may substitute any true clause for "Henry (is) ill on July 28, 1940" and (5) will remain true. Thus, it is true that

(6) "Henry is ill" as spoken by John on July 28, 1940 in W is true in W if and only if the sun is ninety-three million miles from the earth in W.

If (5) gives the meaning of the utterance of "Henry is ill" by stating with truth world-indexed truth conditions of this utterance, then (6) would give the meaning of this utterance for the same reason. But, obviously, "Henry is ill" as used by John does not mean that the sun is ninety-three million miles from the earth. A truth condition sentence that gives this semantic content (or at least gives it up to logical equivalence) must state conditions that obtain in all and only the worlds in which this utterance is true; and this can only be done in terms of truth conditions that are not world-indexed.

The world-indexing of the truth conditions runs into the same problem as does an extensional reading of the biconditional in truth condition sentences that are not world-indexed. It is worthwhile to point this out, since Smart writes that his preferred response to my criticism is to make the biconditional extensional, rather than to read it as intensional and world-index the truth conditions.[22] But if we make extensional the biconditional in

(2) "Henry is ill" is true as spoken by John on July 28, 1940 if and only if Henry (is) ill on July 28, 1940,

then (2) no longer gives the meaning of the utterance of "Henry is ill." If (2), as read extensionally, did give the meaning of the utterance of "Henry is ill," this would be due to the fact that this utterance has the same actual truth value as "Henry (is) ill on July 28, 1940," since the criterion of truth for such extensional truth condition sentences is that the clause after the biconditional has the same actual truth value as the utterance mentioned before the biconditional. But since the utterance of "Henry is ill" also has the same actual truth value as "The sun is ninety-three million miles from the earth," the conclusion that this utterance meant that the sun was ninety-three million miles from the earth would be equally warranted. Only a stricter criterion of truth can make truth condition sentences meaning-giving, namely, the criterion that the clause mentioned after the biconditional is true in all and only the possible worlds in which the mentioned utterance is true. D. H. Mellor makes this point in an intuitively clear manner in *Real Time:*

> For a sentence's truth conditions to give its meaning, its being true in them must be more than a coincidence. Otherwise, so far as truth conditions go, the English sentence "Snow is white" could just as well mean that grass is green, since "snow is white" *is* true and grass *is* green. "Snow is white" is indeed true if and only if grass is green [if the biconditional is read extensionally]. But that of course, is just a coincidence. Even if grass were not green, "Snow is white" would still be true— provided snow was still white. "Snow is white" is not only true in the real world, it would also be true in any other world in which snow was white, and false in any world in which it wasn't. That is really why the sentence means in English what it does, rather than meaning that grass is green. To give meanings, therefore, truth conditions generally have to include imaginary conditions as well as real ones.[23]

Of course some alternate versions of extensional semantics have been developed (e.g., by H. Field, M. Platts, J. McDowell, and the progressively more complicated versions of D. Davidson);[24] but in a well-argued article,[25] Nicholas Asher has shown these versions succumb either to the problem I ascribed to the extensional reading of (2) or to some equally fatal problem. (Further criticisms of extensional semantics can also be found in chapter 5 of Stephen Schiffer's *Remnants of Meaning*.)[26]

Oaklander's objection to an argument I presented in an earlier article[27] may be reduced to Smart's response and thereby rejected for the same reason. Referring to my example of a true 1814 utterance U of "The Battle of Waterloo is present," Oaklander concedes that my example of a possible world in which U is simultaneous with this battle but does not occur in 1814 (where "1814" is read referentially) shows that the truth condition sentence "U is true if and only if the Battle of Waterloo occurs in 1814" is false. In this world (call it W*), U is a member of a different set of simultaneous events, the set directly referred to by "1814*" and in this world, the date-involving truth conditions of U are given (according to Oaklander) by a sentence stating that *the token U "of 'The Battle of Waterloo is present' uttered in 1814* will be true iff the Battle of Waterloo occurs in 1814*."*[28] But this statement is either false or must be read as an extensional or world-indexed truth condition sentence, in which case, it is semantically inadequate for the same reason as Smart's response. This statement is false if "iff" is read as an intensional, rather than extensional, biconditional and the statement is not a world-indexing one, since there is some other world W** in which U is simultaneous with the Battle of Waterloo (and thus is true) but is not a member of the set 1814*. But if Oaklander's truth condition statement is true, then the "iff" is extensional or the statement is implicitly world-indexing (so that "in W*" belongs after each occurrence of "in 1814*"); and in either case, the statement runs into the same problems as Smart's modified theory.

Oaklander offers a second argument, but it runs into the same problems. If we call the set of events directly designated by "1814" the set B, such that U is actually a member of B but is not a member of B in some other possible worlds, then it follows, according to Oaklander, that "if U does not occur in B, then U cannot be a true 1814 token of 'The Battle of Waterloo is present'. Thus, . . . Smith cannot claim that (an 1814 token) U is true even if the Battle of Waterloo does not occur in 1814."[29] Oaklander's remarks may be understood to be implicitly world-indexed, as meaning, "If U does not occur in B in W*, then U does not possess in W* both the properties of *being true* and *occurring in 1814*. Thus, Smith cannot claim that in W*, U is both true and occurs in 1814 if the Battle of Waterloo does not occur in 1814 in W*." Understood this way, Oaklander's remarks are true but irrelevant, since the claims that I in fact made were not about the world-indexed truth conditions of U but about its non-world-indexed and nonextensional truth conditions; and these claims *are* true, since U is true in worlds in which the Battle of Waterloo does not occur in 1814. Oaklander's remarks are also true if they are interpreted as claims about the extensional truth conditions of U; but this is again irrelevant, since these claims are consistent with my claims about the intensional and non-world-indexed truth conditions of U. Consequently, Oaklander, like Smart, fails to resolve the vital problem with the date-sentence theory of A-sentences, that date-sentences fail to *express* (contra the old tenseless theory) or *give* (contra the new tenseless theory) the semantic content (meaning) of A-sentence-tokens, since they do not state the intensional and non-world-indexed truth conditions of A-sentence-tokens. The intensional and non-world-indexed truth conditions of these tokens, stated in tenseless terms, are instead *token-reflexive* truth conditions of the form "T (is) related by B to E," where T stands for the A-sentence-token; B, for a B-relation of simultaneity, earlier, or later; and E for the event that the sentence-token is about. In chapter 4 I shall show that the fact that A-sentence-tokens have tenseless token-reflexive truth conditions, as tenseless date-sentences do not, and the fact that tenseless date-sentences have date-involving truth conditions, as A-sentence-tokens do not, is ultimately explained *only* by the hypothesis that A-sentence-tokens, but not tenseless date-sentences, ascribe A-properties.

But we are still a long way from being able to argue for the soundness of this explanation, since there are other detenser arguments pertaining to other semantic properties of A-sentence-tokens that might seem to show that these sentence-tokens, despite the aforementioned differences from tenseless date-sentences, have the same semantic contents as the latter. These arguments will be considered in the following three sections.

2.3 The Confirmation Conditions of A-Sentence-Tokens and Tenseless Date-Sentences

Two sentences or sentence-tokens have the same confirmation conditions if and only if they are confirmed or disconfirmed to the same degree by observations of the same sort. That A-sentence-tokens have confirmation conditions different from their corresponding date-sentences follows immediately from their difference in truth conditions. Consider "Henry Jones of Lee St., Tulsa is ill" as tokened on some occasion. An observation that confirms this token as true is an observation that Henry's illness occurs simultaneously with the token. However, no observation of a similar sort could confirm any token of "Henry Jones of Lee St., Tulsa (is) ill on July 28, 1940." To observe that Henry's illness is simultaneous with some token of this sentence neither confirms nor disconfirms this token; for suppose that some token of this sentence is observed to be simultaneous with Henry's illness but that the date of Henry's illness is unknown; we would then not be in possession of enough information to determine the truth value of the token. Generalizing, we can say that the first difference in confirmation conditions between A-sentence-tokens and date-sentences and their tokens is that observations of B-relations between the event reported and the reporting sentence-token are *sufficient* to confirm the A-sentence-tokens but *insufficient* to confirm the date-sentence-tokens.

The second difference is that observations of the B-relation between the event reported and the date are *sufficient* to confirm the date-sentence but are *insufficient* to confirm the A-sentence-token. Each token of "Henry Jones of Lee St., Tulsa (is) ill on July 28, 1940" is confirmed by the observation that Henry's illness occurs on July 28, 1940; but no such observation confirms the A-sentence-token. For this observation to count toward the confirmation of the A-token, I must possess, *in addition,* observational evidence that the A-token occurs on July 28, 1940, that is, that the A-token is simultaneous with Henry's illness. And this is tantamount to saying that the observational evidence that Henry's illness occurs on July 28, 1940 is not *by itself* sufficient to confirm the A-sentence-token.

The difference in confirmation conditions between an A-sentence-token and its corresponding tenseless date-sentence become especially acute in the case of *A-date-sentence-tokens* and their corresponding tenseless date-sentences. An example of an A-date-sentence is "It is noon" or "Today is February 5." The confirmation conditions of these A-date-sentences and their alleged tenseless translations will be my particular focus in this section. I shall first critically examine Grünbaum's theory of A-date-sentence-tokens, then William Lycan's theory, David Kaplan's and Nathan Salmon's theories, and finally Michelle Beer's theory.

A. Adolf Grünbaum's Theory of A-Date-Sentence-Tokens

Consider a token T_1 of "10:30 A.M. E.S.T., February 26, 1986 is present" that occurs on this date. T_1 is a logically contingent sentence-token that is confirmed by observations that it is simultaneous with 10:30 (etc.). This point may be obvious, but it is worth driving home. It is manifest that in order to determine whether T_1 is true or false it is not sufficient that I simply *understand* T_1. I must also be in possession of observational evidence about T_1's B-relation to 10:30. For instance, if somebody utters T_1, I cannot simply nod my head and say, "Of course that's true." I must also look at my watch to see if T_1 is borne out by the empirical evidence. But consider now the tenseless date-sentence that corresponds to T_1: "10:30 A.M., E.S.T., February 26, 1986 (is) at 10:30 A.M., E.S.T., February 26, 1986." This sentence is logically necessary; the date described by the first date-expression is necessarily simultaneous with the date described by the second date-expression, since they are one and the same date. Consequently, I know this sentence to be true independently of any observational evidence and simply through understanding it. This is prima facie evidence that the indirect reference date theory is wrong in stating that T_1 and the date-sentence express the same proposition.

It is worth making clear that according to the indirect reference date theory, the aforementioned tenseless date-sentence *is* the one that corresponds to T_1. It is a tenet of this theory that the A-predicate "present" conjoined with the A-copula "is" refer indirectly to the date on which they are tokened. If "x is present" is uttered at 10:30, then "is present" expresses the propositional definite description that refers to this date. This entails that in the case of "10:30 (etc.) is present," the item referred to by the rest of the sentence is a date, which entails that the date is referred to twice, once by the date-expression and once by "is present." What the token T_1 of this sentence asserts, according to the date theory, is that the date referred to by the date-expression is simultaneous with the date referred to by "is present," that is, that 10:30 (etc.) is simultaneous with 10:30 (etc.). And thus the logic of the indirect reference date theory drives us to the seeming paradox that a logically contingent sentence-token expresses the same proposition as a logically necessary sentence.

A defender of the indirect reference date theory might try to solve this problem by arguing that the A-date-sentences or their tokens are no less logically necessary than their corresponding tenseless date-sentences. Adolf Grünbaum, taking this approach, writes that the tenser

> must construe the assertion "It is 3 P.M., E.S.T., now" as claiming *non-trivially* that when the clock strikes 3 P.M. on the day in question, this clock event and all the events simultaneous with it intrinsically have the unanalyzable property of nowness or presentness. But I am totally at a loss to see that anything non-trivial can possibly be asserted by the claim that at 3 P.M. nowness (presentness) inheres in the events of 3 P.M. For all I am able to discern here is that the events of 3 P.M. are indeed those of 3 P.M. on the day in question.[30]

Grünbaum's argument is based on a confusion between

(1) It is 3:00 P.M. now

and

(2) At 3:00 P.M. nowness inheres in the events of 3:00 P.M.

Grünbaum believes that (1) and (2) make the same assertion and that since (2) is trivial, (1) is also trivial. But manifestly, "It is 3:00 P.M. now" and "At 3:00 P.M. nowness inheres in the events of 3 P.M." make different assertions. Tokens of the first sentence are in some cases false, namely, when they occur at times earlier or later than 3:00 P.M. Surely it is an obvious fact about English that if I utter "It is 3:00 P.M. now" at 5:00 P.M., I have uttered something false. If so, "It is 3:00 P.M. now" cannot be a *trivial truth;* and surely, it is equally obvious that if I utter "At 3:00 P.M. nowness inheres in the events of 3:00 P.M.," I say something true no matter when I say it, be it 3:00 P.M. or 5:00 P.M. This sentence is no less trivially true than "At

3:00 P.M. pastness inheres in the events of 3:00 P.M." is trivially false. Consequently, one cannot infer from the triviality of sentences like (2) to the triviality of
sentences like (1); and the attempt made by Grünbaum to show that A-date-
sentences like (1) are trivially true must be counted a failure.

B. William Lycan's Theory

The particular problem we found with the indirect reference theory of A-date-
sentences is avoided by the direct reference theory of these sentences, but a new and
related problem arises to take its place. The direct reference theory holds that
expressions such as "now" refer directly to the date on which they are tokened,
without the intermediary of a propositional definite description of this date. This
seems to demand that the corresponding tenseless date-sentence also possess a
phrase that directly refers to a date and that thereby captures the A-words' direct
reference to the date. This demand was first met in William Lycan's account of
these tenseless date-sentences. Lycan recognizes that what is expressed by an utterance of "It is 4:30 P.M." at 4:30 P.M., March 1, 1967, in Lincoln, Nebraska is
logically contingent; but he argues that what is expressed by this utterance is also
expressible by "It is 4:30 P.M., March 1, 1967, in Lincoln, Nebraska at 4:30 P.M.,
March 1, 1967, in Lincoln, Nebraska," a sentence that Lycan abbreviates as "It is
4:30 at 4:30." Lycan continues:

> There is a natural way of construing "It is 4:30 at 4:30" according to which the
> sentence is contingent and non-trivial. . . . Let us construe the second occurrence
> of "4:30" as a fused referring expression, serving to pick out a particular moment
> (the moment at which the speaker happened to speak, under whatever designation);
> the whole sentence may then be read as a singular predication having the second
> occurrence of "4:30" as its subject. Thus, one who utters the sentence would be
> predicating something of the moment 4:30 (however referred to), viz., the property
> of its-having-been-4:30-at that moment.[31]

Lycan later indicates that the property of its-having-been-4:30-at that moment is
more fully specifiable as some such property as *being four-and-one-half hours later
than the moment at which the sun reaches its highest point in the sky*. He also
corrects the impression that might be conveyed by this passage to the effect that the
second occurrence of "4:30" refers to a *different* moment at each different moment it
is used, an impression conveyable by his assertion that this expression serves to pick
out "the moment at which the speaker happened to speak." Lycan believes the
contrary is true; since "4:30" "in fact denotes 4:30 in our language, . . . in saying
'It is 4:30 at 4:30,' . . . we know that we *are* talking about the moment 4:30, and
the statement therein made will not fluctuate from speaker to speaker."[32] Thus,
"4:30" refers to the same time, 4:30, at each different time that "It is 4:30 at 4:30" is
uttered. (Recall that "4:30" is an abbreviation for "4:30 P.M., March 1, 1967, in
Lincoln, Nebraska.") Lycan's remark that "4:30" serves to pick out "the moment at
which the speaker happened to speak" must accordingly be interpreted as offering
an example of a speaker who happened to speak *at 4:30* and, for this reason,
referred to the moment of his speaking.

The core of Lycan's interpretation of "It is 4:30 at 4:30" as logically contingent lies in his remark that the second "4:30" is a "fused referring expression." Lycan hints that he means by this that this "4:30" is a referentially, rather than attributively, used definite description, in Donellan's sense.[33] Lycan does not elaborate upon this hint; but his meaning seems plain, namely, that if "4:30" is used referentially, it directly refers to the date, 4:30, and if attributively, it indirectly refers to this date by expressing the propositional definite description, *whatever set of events (or moment) is four-and-one-half hours later than the sun's zenith.* If "4:30" is used referentially, then the sentence "It is 4:30 at 4:30" will express the *de re* proposition consisting of the ordered pair of the date 4:30 and the property *being four-and-one-half hours later than the sun's zenith:*

(3) ⟨4:30, the property of being four-and-one-half hours later than the sun's zenith⟩.

If (3) is the proposition expressed by "It is 4:30 at 4:30," then this sentence is logically contingent. This is because the date 4:30 (4:30 P.M., March 1, 1967, in Lincoln, Nebraska) is a certain set S_1 of events or moment M_1 that *in fact* has the property of being four and one-half hours later than the sun's zenith (on this day) and does not have this property in some other possible worlds.

This *de re* translation of "It is 4:30" seems to solve the problem confronting the *de dicto* translation of A-date-sentence-tokens. But in fact, two new problems emerge. One is that the confirmation conditions of the *de re* reading of "It is 4:30 at 4:30" are different from those of the token T_1 of "It is 4:30." Observational evidence that confirms the *de re* reading of "It is 4:30 at 4:30" is evidence that the set of events S_1 or moment M_1 possesses the property of *being four-and-one-half hours later than the sun's zenith.* But this evidence is insufficient to confirm the token T_1 of "It is 4:30" that occurs at 4:30. If I did not possess observational evidence that T_1 is simultaneous with 4:30 and only possessed evidence that S_1 or M_1 is four-and-one-half hours later than the sun's zenith, I would not be in a position to determine whether T_1 was true or false. In order to confirm T_1, I must possess, in addition to the observations that confirm "It is 4:30 at 4:30," observations of a *different sort, observations of the B-relation between the sentence-token and the time S_1 or M_1.* (S_1 or M_1 are picked out or identified observationally in terms of events that belong to them.)

A second problem is that A-date-sentence-tokens whose date-expressions are *directly referential* are confirmed by certain observations and disconfirmed by others, but their tenseless *de re* translations are trivially true or false. In other words, a problem similar to the one confronting the indirect reference date theory arises for the direct reference date theory, albeit for a different reason. The problem for the indirect reference date theory concerns A-date-sentence-tokens wherein the date-description is used *attributively.* The direct reference date theory is able to provide logically contingent translations of *these* A-date-sentence-tokens by arguing that the A-expressions in the A-date-sentence-tokens are translated by tenseless date-descriptions used referentially, rather than attributively. But there are other A-date-sentence-tokens in which the date-expressions are used *referentially.* Suppose I utter "It is 4:30" at 4:30 and use "4:30" referentially to pick out the set of events S_1, the set that is, in fact, four-and-one-half hours later than the sun's zenith (on March 1, 1967 in Lincoln, Nebraska). My utterance is not trivially true but needs to be confirmed by the observation that it is in fact a member of S_1. Somebody who is not in possession of this information would not be able to evaluate the truth of my utterance. But if the direct reference date theory is correct and my 4:30 utterance of the present tense "It is," is to be translated by the *de re* date-description "It is 4:30 at . . ." (where the copula is, of course, tenseless), then the result is the *tautology* "It is 4:30 at 4:30." This is a tautology because both occurrences of "4:30" are referentially used definite descriptions that refer directly to S_1, such that this date-sentence is synonymous with "It is S_1 at S_1."

Let us highlight this difference in confirmation conditions between these referentially used A-date-sentence-tokens and their alleged *de re* translations. Suppose this conversation occurs at S_1, at 4:30, and "4:30" is used referentially by both conversationalists:

> A: Is it 4:30 yet?
>
> B: Yes, it is 4:30.
>
> A: I see. Thanks for the information.

Clearly, B's answer does not state what "S_1 occurs at S_1" states; for A already knew that trivial fact and would hardly thank B for conveying it to her. Rather, she learned some observationally derived information that she was not in possession of beforehand, information that (in tenseless terms) S_1 *is occurring simultaneously with the conversation.*

C. David Kaplan's and Nathan Salmon's Implicit Theories

But these arguments would not trouble all direct reference date-theorists, since some of them believe they have a ready explanation of these differences in confirmation conditions. David Kaplan, for example, would acknowledge that a use of a sentence containing both a temporal indexical and a *de re* date-description imparts a posteriori and observationally verifiable information but would insist that this is consistent with this sentence use expressing the same trivially true proposition as the corresponding tenseless date-sentence. Kaplan would point out that the empirical information supplied by the sentence-token that contains the token of the temporal indexical is not supplied by the semantic *content* of this token—the proposition it expresses—but by the pertinent semantic *character,* specifically, by the rule of use that determines the reference of the temporal indexical on each occasion of its use. Using the phrase "cognitive significance" to refer to the informativeness of sentence-tokens, Kaplan writes, "We identify object of thought [the proposition or propositional constituent expressed] with content and the cognitive significance with character."[34] Consider John's utterance at noon of "It is now noon." Kaplan would say that John's utterance, call it U, expresses the same tautologous *de re* proposition that is expressed by "Noon (is) at noon" (where "noon" is an abbreviation of some complete *de re* date-description) but that U has a different "cognitive significance" or information value than the tenseless utterance by virtue of being a token of a sentence that has a different "character" or rule of use. The sentence "It is now noon" is governed by the rule that *"It is now noon" refers directly to the time of its use and asserts of it that it (is) simultaneous with noon,* and the tenseless sentence is not governed by this rule. By virtue of John and his addressee's grasping this rule of use, they find U to have a "cognitive significance" and to convey empirically confirmable information that the use of the tenseless sentence does not.

But I believe this theory does not succeed in explaining the cognitive significance of John's utterance of U. If I grasp the character

(4) The sentence "It is now noon" refers directly to the time of its use and asserts of it that (is) simultaneous with noon,

that does *not* provide me with the empirical information that John's utterance U of "It is now noon" gives me, for (4) does not entail that U occurs at noon. Indeed, (4) entails no information about U or about noon, since (4) is a general statement that abstracts from particular events and times. In fact, (4) does not even have a truth value, since it is a conventional stipulation and stipulations are neither true nor false. This remark ties in with a criticism that Hector-Neri Castañeda and Palle Yourgrau have made of Kaplan's theory of cognitive significance, that his characters are general formulae for determining the semantic contents of expressions and that the cognitive significance of uses of indexical-containing sentences are truth-valued pieces of information about individual times, places, or persons.[35]

But it seems Kaplan's theory provides the key to constructing a theory of cognitive significance as something other than the semantic content of sentence-tokens, for characters may be taken as proper parts of this significance. There is a distinction made by Nathan Salmon, Howard Wettstein, Michael Tye, Raymond Bradley, Norman Swartz, Tom McKay, and R. M. Sainsbury, between the *semantic information* of a sentence-token, (corresponding to Kaplan's semantic content) and the *pragmatically imparted information* of a sentence-token (a wider notion than Kaplan's character but which may include his characters as proper parts).[36] Although these authors have not applied this distinction to A-sentence-tokens that include *de re* date-descriptions, this application seems relatively straightforward. Let us suppose the cognitive significance of some sentence-tokens may include a character and the subsumption of the sentence-token under the character; that is, the cognitive significance is the information that the utterance is an individual instance of the expression-type mentioned in the statement of the character. If we adopt this theory, we will say that (4), in conjunction with the subsumption

(5) John's utterance U (is) a correctly used token of "It is now noon,"

provides the cognitive significance of U; for (4) and (5) jointly entail

(6) U pragmatically imparts the information that it (is) simultaneous with noon.

With this conception of cognitive significance we have, finally, some success for the tenseless date-theory; for (4)–(6) meet a necessary condition of stating the cognitive significance of U, namely, they entail the verifiable information about U's token-reflexive B-property, *being simultaneous with noon*. This theory of cognitive significance thus constitutes an advance over the theory of Kaplan and the previously discussed theories of Lycan and Grünbaum. But I would now like to introduce a new consideration that shows that (4)–(6) fail to meet another necessary condition of stating the cognitive significance of U. This consideration can be presented most exactly in terms of the most completely developed of the aforementioned post-Kaplan theories of cognitive significance, the version developed by Nathan Salmon in *Frege's Puzzle*.

Salmon holds that A-sentence-tokens directly refer to dates and do not ascribe A-properties,[37] which implies that the token U of "It is now noon" expresses the logically necessary proposition, "Noon (is) at noon." This is the semantically imparted information of U; but its pragmatically imparted information involves a *propositional guise* (not to be confused with Castañeda's propositional guises), which is a "manner of presentation" or "mode of appearance" of the proposition. Salmon writes that this distinction allows that "the semantically encoded information may be knowable a priori even when the sentence's pragmatic impartations are knowable only a posteriori."[38] If we apply this theory to our example, we may say that among the pragmatic impartations of "It is now noon" is that this sentence is used simultaneously with noon, an impartation that can be confirmed to correspond to the facts only through empirical observations. If we take this imparted information to be the guise x by means of which the singular proposition *noon is at noon* is grasped, we may apply Salmon's schema for belief to our example. The schema[39] is

(7) "A believes p" may be analyzed as ($\exists$x) [A grasps p by means of x & BEL (A,p,x)],

where p is the proposition, x the propositional guise, and BEL a triadic belief relation. Applied to our example, (7) may be given the interpretation:

(8) "Alex believes the proposition expressed by John's utterance U of 'It is now noon'" may be analyzed as follows: Alex grasps the proposition *that noon is at noon* by means of the propositional guise *being the proposition that ascribes self-simultaneity to the time with which U is simultaneous;* and Alex stands in the triadic belief relation BEL to *that noon is at noon [such that this is the proposition that ascribes self-simultaneity to the time with which U is simultaneous]*.

Every verb and copula in (8) is tenseless. My argument is that (8) is true only if

(9) Alex believes *that noon is at noon [such that this is the proposition that ascribes self-simultaneity to the time with which U is simultaneous]* is now true

entails

> (10) Alex believes that what John semantically and pragmatically conveyed to him by his noon utterance of "It is now noon" is now true.

But (9) does not entail (10), since it is now 1:30 and Alex now believes (correctly) *that noon is at noon [such that this is the proposition that ascribes self-simultaneity to the time with which U is simultaneous]* is now true, but he nonetheless now believes (also correctly) that what John's noon utterance of "It is now noon" conveyed is now false. Of course, Alex now believes that what U conveyed *was* true when U was uttered; but this is consistent with his believing that what U conveyed, unlike the tenseless-proposition-cum-tenseless-guise is no longer true. If one prefers these facts stated in terms of beliefs in or about U, it may be said that Alex can consistently believe that it is now true that John's utterance U of "It is now noon" is simultaneous with noon and that noon is at noon without believing that U is now true, which entails that the belief in the simultaneity of noon with itself and with U is not logically equivalent to the belief in U.

If it is objected that truth values of utterances cannot change over time and that the assumption that they do presupposes the tensed theory of time, I would respond by denying both claims. Regarding the latter, change of truth value over time does not presuppose the tensed theory, since these changes can be described in purely tenseless terms and in terms of B-relations. For example, it can be said (with every verb and copula being tenseless),

> (11) At 1:30, Alex believes that what U conveys is *then* (at 1:30) false but is true *earlier,* at noon

and

> (12) At noon, what U conveys is true and at 1:30 what U conveys is false.

"What U conveys" in (11) and (12) means "what U conveys at noon." Now (11) and (12) mention no A-properties and contain no tenses, so they can be consistently represented as descriptions of some of the tenseless semantic properties of A-sentence-tokens. Furthermore, the fact that the thesis *some truth vehicles change their truth value over time* does not presuppose the tensed theory of time is further evinced (or suggested) by its endorsement by some proponents of the tenseless theory of time. Both Kaplan[40] and Ernest Sosa[41] hold that propositions do not ascribe A-properties but nonetheless maintain that some propositions change their truth value over time. If it is insisted, nonetheless, that truth-bearers *cannot* change their truth value, I would point out that this claim is a theoretical postulate and cannot legitimately be said to be a description of the linguistic data that the tenser and detenser are both trying to explain. To use Castañeda's terminology, (11) and (12) belong to *protophilosophy* (data description), rather than *diaphilosophy* (data explanation and theory building), whereas "Truth-bearers cannot change their truth value" is a diaphilosophical statement.[42] If it is denied that (11) and (12) are statements in protophilosophy, I respond that they and their tensed analogues reflect "how we ordinarily talk" or "what we would ordinarily say" and that this can be borne out by observation. For example, we ordinarily say such things as "What John stated when he said 'It is now noon' *was* true when he said it but is *now* false." I believe, accordingly, that it is fair to conclude that the Salmon-based theory of the cognitive significance of A-date-sentences, although it meets one necessary condition of explaining their significance (i.e., it explains the information they convey about their B-relatedness to dates), fails to meet the second necessary condition, namely, that is explain their cognitive significance's change in truth value over time.

However, I do not want to leave the impression that my argument for the tensed theory of time crucially requires the premise that truth vehicles can change their truth value over time, since my argument could then be too easily dismissed by the die-hard supporters of unchangeable truth values. Accordingly, I shall provide, against the detenser theory I constructed in terms of Salmon's belief schema, an argument that uses premises logically independent of the premise that U changes its truth value. I shall show that this theory fails to meet a third necessary condition of stating the cognitive significance of U, namely, that it explain the logical equivalence of the cognitive significance of U with the cognitive significance of a simultaneous token of a synonymous sentence. If a person A believes an utterance U at noon of "It is now noon" and also believes an utterance V of the synonymous sentence "It is noon now" that occurs simultaneously, then she has the third (perhaps implicit) belief that the objects of her first two beliefs are logically equivalent. Relative to the theory we are considering, "beliefs" designates the three-termed relation BEL that Salmon introduces. But if the Salmon-based theory is true, the object of A's belief regarding the utterance U is the proposition-cum-propositional-guise

(13) that noon is at noon [such that this is the proposition that ascribes self-simultaneity to the time with which U is simultaneous],

and the object of A's belief regarding V is

(14) that noon is at noon [such that this is the proposition that ascribes self-simultaneity to the time with which V is simultaneous].

But (13) and (14) are not logically equivalent, since the fact that the proposition *that noon is at noon* has the property of being about the time with which U is simultaneous does not entail that it has the property of being about the time with which V is simultaneous (somebody could have uttered U without anybody uttering V). Therefore, this Salmon-based theory fails to meet this third necessary condition of adequacy that a theory of cognitive significance must meet, namely, that the cognitive significances of simultaneous utterances of synonymous A-sentences in exactly the same circumstances be logically equivalent. Further criticisms of this Salmon-based theory of the *token-reflexive* nature of the cognitive significance of A-sentence-tokens are implicit in the criticisms I shall put forth in chapter 3, where I discuss the token-reflexive theory of A-sentences.

D. Michelle Beer's Theory

I shall critically examine other detenser theories of cognitive significance and of the semantic/pragmatic distinction in the following two sections, but I should like to close this section by saying a few words about a detenser theory of A-date-sentences that differs from all of those considered so far. I have in mind the theory of A-date-sentences that belongs to Michelle Beer's new tenseless theory of time.

Beer argues that A-sentence-tokens do not ascribe A-properties (conceived as irreducible to B-relations), even though these tokens express propositions different from those of B-sentences. Thus, unlike all the date-theorists considered so far in this section, she *admits* the thesis for which I have been arguing, that an utterance of an A-date-sentence expresses a proposition different from that of the corresponding tenseless date-sentence. For example, Beer avers that an utterance at t_7 of "It is t_7 now" expresses a proposition different from an utterance of "It is t_7 at t_7," such that the two utterances "differ in sense."[43] Beer then claims that this fact is consistent with the A-utterance and the B-utterance's being such that "they refer to the same subjects or relata [viz., t_7 and t_7] and ascribe the same relation between them [simultaneity]" and concludes from this that the A-utterance does not ascribe an A-property (conceived as irreducible to a B-relation).[44] But this argument, although it has convinced Richard Gale[45] and L. Nathan Oaklander,[46] seems to me unsound. Note, to begin with, that Beer's premise that the A and B utterances differ in sense but refer to the same relata and ascribe the same relation between them is ambiguous, yielding two interpretations. On one reading, it means that the two utterances differ in sense but refer in the same way(s) to all and only the same items and ascribe to these items the same n-adic properties. On this reading, the premise is self-contradictory, since differences in sense entail differences in referents, or ways of referring, or n-adic properties ascribed to the referents. On a second reading, the premise is self-consistent and means that two utterances both refer to t_7 and ascribe to it *being simultaneous with itself,* even though (by virtue of their "difference in sense") one of them refers to something the other does not or in a way the other does not or ascribes some n-adic property the other does not. But on this reading, the conclusion that the A-utterance does not ascribe an irreducible A-property does not follow, since the premise is consistent with the A-utterance differing in sense from the B-utterance by virtue of ascribing an irreducible A-property to t_7 or referring to t_7 in a way that involves the propositional definite description *the time that exemplifies the irreducible A-property of presentness.*

It is possible that a detenser might try to defend Beer's argument along the following lines. The detenser will note that at one point Beer mentions the relevant difference in sense between the A-utterance and the B-utterance. She says they differ in sense in that "the proposition expressed by now [at t_7] using 't_7 is now' informs us that t_7 is present or now, whereas that expressed by 't_7 is (tenselessly) simultaneous with t_7' does not."[47] This statement is *not* consistent with the tensed theory of time, it will be said, since Beer holds that "an event's having an A-determination—its being past, present, or future—is identical with that event's bearing a temporal relation [a B-relation] to some moment of time."[48]

But this line of defense is plainly untenable. For one thing, it does not avoid the first horn of the aforementioned dilemma, the contradiction that the A and B utterances differ in sense but refer in the same ways to the same items and ascribe to them the same n-adic properties. If t_7's being present is identical with t_7's being simultaneous with t_7, then the A-utterance, by virtue of ascribing presentness to t_7 or referring to t_7 as present, does not ascribe to t_7 any n-adic property or refer to t_7 in any way that the B utterance does not, which is inconsistent with their differing in sense. How could the information that t_7 is present be *different* from the information that t_7 is simultaneous with t_7 if t_7's being present is *identical* with its being simultaneous with t_7? Second, the proffered line of defense avoids the second horn of the dilemma, that the difference in sense is consistent with the A-utterance's ascribing an irreducible A-property, only at the price of being overtly question-begging, since the thesis that "t_7's being present is identical with t_7's being simultaneous with t_7" now becomes a *premise,* rather than the conclusion of the argument against the tensed theory of time. I conclude that Beer's new tenseless theory of time—Gale's and Oaklander's endorsement notwithstanding—fails to rescue the tenseless date theory from the quandary posed by A-date-sentences. (Further criticisms of Beer's theory are presented elsewhere.)[49]

2.4 The Entailment Relations of A-Sentence-Tokens and Tenseless Date-Sentences

I shall discuss in this section two sets of entailment relations that are possessed by A-sentence-tokens but not by their corresponding tenseless date-sentences. The first set concerns entailments possessed by some declarative A-sentence-tokens, the second, entailments possessed by certain modal A-sentence-tokens.

The account of the truth conditions and confirmation conditions of A-sentence-tokens in the last two sections provides the key for uncovering the first set of entailment relations to be discussed. In terms of our "Henry is ill" example, we may say that each token of this sentence entails

   (1) The illness of Henry (is) simultaneous with some sentence-token about this illness;

for each token of this A-sentence is true only if it is simultaneous with Henry's illness. However, (1) is not entailed by "Henry (is) ill on July 28, 1940" or any one of its tokens; for each token of this sentence could be true even if no tensed or tenseless token about Henry's illness occurred simultaneously with his illness.

Nelson Goodman believes that an argument to the effect that A-tokens do, and tenseless date-tokens do not, convey information about the simultaneity of the event with the tokens is not sufficient to show that A-tokens cannot be translated by the date-tokens. Goodman calls tokens like A-tokens "indicators"; an indicator is any token whose replica has a different denotation. He writes:

> We may seek a translation that contains no name of the indicator itself, but rather another name for what the indicator names. Thus a certain "here" is translated by any "Philadelphia" and a certain "ran" is translated by any

"runs [tenseless] on Jan. 7, 1948 at noon E.S.T."

Against such translations, it is sometimes urged that they do not really convey the content of the originals. A spoken

"Randy is running now"

tells us that the action takes place at the very moment of speaking, while a

"Randy runs [tenseless] on October 17, 1948 at 10 P.M., E.S.T."

does not tell us that the action takes place simultaneously with either utterance unless we know in addition that the time of the utterance is October 17, 1948 at 10 P.M. E.S.T. Since—the argument runs—we recognize the tenseless token[50] as a translation of the tensed one only in the light of outside knowledge, we have here no genuine translation at all. But this seems to be no more cogent than would the parallel argument that "L'Angleterre" is not a genuine translation of "England" because we recognize it as a translation only if we know that L'Angleterre is England.[51]

Goodman's point is that if a token of "Randy is running now" occurs on October 17, 1948 at 10:00 P.M., E.S.T., then it is translated by a token of "Randy (runs) on October 17, 1948 at 10:00 P.M., E.S.T.," which we will suppose occurs on the same date. If we know that the latter token occurs on the same date as the action, we know that it is simultaneous with the action. Consequently, we will be in possession of the same information in regard to the tenseless token as is conveyed by the tensed token, namely, that the action takes place at the very moment the token occurs.

Contrary to Goodman's contention, the knowledge that the tenseless token is simultaneous with the action *is* an illegimate "outside knowledge" and thus renders the translation spurious. A necessary condition of some token T_1's genuinely translating another token T_2 is that T_1 can be known to translate T_2 solely on the basis of a knowledge of the rules of usage of the tokens in the class of tokens that includes T_1 and all tokens similar to T_1 and of the tokens in the class of tokens that includes T_2 and its replicas. The knowledge involved here is not "outside knowledge" but is just that knowledge needed for any translation. Now the translation of some token of "England" by some token of "L'Angleterre" meets this condition, for I recognize the token of "L'Angleterre" to be a translation of the token of "England" solely on the basis of my knowledge that all tokens in the class of tokens that includes the English token and all tokens similar to the English token are used to refer to the same nation as are all tokens in the class of tokens that includes the French token and its replicas. The translation of the token of "Randy is running now" that occurs on October 17, 1948 at 10:00 P.M., E.S.T. by the token of "Randy (runs) on October 17, 1948 at 10:00 p.m., E.S.T." that occurs on the same date does not meet this condition; for this translation requires a knowledge of a different sort. This is a consequence of the fact that all tokens in the class of tokens including only the tensed token and its replicas have different rules of usage from the tokens in the class of tokens including only the tenseless token and its replicas. The tensed token and its replicas are used to refer to events that are simultaneous with the tokens, but the tenseless token and its replicas are not so used. The latter tokens are, instead, used to refer to events that occur on October 17, 1948 at 10:00 P.M., E.S.T. Consequently, I know by virtue of understanding the rules of usage of the tensed token and its replicas that the tensed token occurs simultaneously with the action (supposing the tensed token is true); but I cannot know that the tenseless token (supposing it also is true) is simultaneous with the action simply by virtue of understanding the rules of usage of this token and its replicas. Instead, I must resort to outside knowledge, that is, observational evidence that somebody uttered the tenseless token simultaneously with Randy's running on October 17, 1948 at 10:00 P.M., E.S.T. In a word, the tensed token, by virtue of its rules of usage, *entails* that the action is simultaneous with the token; and the tenseless token does not. The two tokens are logically nonequivalent and cannot genuinely translate each other.

Some date-theorists, such as Steven Boer and William Lycan, might respond to my argument by asserting that the difference between the A-token and the B-token is not semantic but pragmatic. In their article "Who, Me?" Boer and Lycan claim:

> "Now" is simply a purely referential or rigid designator whose referent is a moment in time, and there is a pragmatic rule to the effect that a token of "now" always refers to the moment of its utterance. . . . There is nothing special about the semantic content of the propositions expressed [by the A-sentence-tokens that distinguishes them from the propositions expressed by the corresponding tenseless date-sentences]; all that is distinctive are the pragmatic rules that compute the indexical terms' denotata.[52]

Although the main argument Boer and Lycan present for this semantic/pragmatic distinction concerns first-person indexicals, a similar argument can be constructed for temporal indexicals. The argument would be that a token of "Randy is running now" does not *entail* that there is some token simultaneous with Randy's running but, instead, that by virtue of its pragmatic rules, it *implicates* this (in Grice's sense). It follows from this that the aforementioned difference between this A-token and the corresponding B-token is not a difference in the entailment relations of the two tokens and, therefore, that this difference is compatible with the two tokens' being logically equivalent.

But this argument is unsound, since a necessary condition of a sentence being a Gricean implicature, rather than an entailment, is that its falsity is consistent with the truth of the sentence or sentence-token that implicates it. But there is no possible world in which the token of "Randy is running now" is true and in which

(2) There is some sentence-token about Randy's running that (is) simultaneous with Randy's running

is false.

The date-theorist might respond by adopting a more liberal conception of the pragmatic/semantic distinction than Boer and Lycan's and assert that *some entailments are pragmatic* and that difference in pragmatic entailments is consistent with having the same semantic content (expressing the same proposition). Semantic entailments, it will be said, are due to the proposition expressed by the sentence-token, and pragmatic entailments are not. If a token T_1 expresses a proposition P_1 that entails, and is entailed by, the proposition P_2 expressed by the token T_2 but T_1 entails something T_2 does not, then T_1 and T_2 are *semantically* logically equivalent but *pragmatically* logically nonequivalent. The date-theorist will say that semantic, rather than pragmatic, logical equivalence is a necessary condition for translation in the relevant sense (where "translates" means "expresses the same proposition as"). The date-theorist will insist that the A-token's entailment of its simultaneity with the event it is about is pragmatic and therefore consistent with the A-token's semantic logical equivalence with the B-token.

There are two ways I could respond to this objection. First, I could argue that the A-token's entailment of its simultaneity with the event it is about demands that it be interpreted as expressing a different proposition from that of the date-sentence-token. Second, I could assume for the sake of argument the date-theorist's claim that the entailment in question is merely pragmatic and consistent with the semantic logical equivalence of A- and B-tokens but go on to show that there are additional logical nonequivalences that are clearly *semantic,* even by the date-theorist's own standards. This second line of defense will be easier to deploy, since it appeals to principles the date-theorist accepts; and since it is sufficient to refute the date-theorist, I shall use only it in what follows.

I will concentrate on modal sentences and use them to show that many A-sentence-tokens are so far from being semantically logically equivalent to the corresponding tenseless date-sentences or tokens that they are not even *materially* equivalent to them.

Consider a token T_3 of

(3) Christ might not have been born during the present year

that occurs during the year of Christ's birth, 1 A.D. T_3 is true when it occurs, which implies that the proposition it expresses is true at least at this time. But T_3 does not entail its corresponding *de dicto* date-sentence,

(4) Christ might not have been born during whatever year includes Christ's birth,

and the proposition expressed by T_3 does not entail the proposition expressed by (4); for (4) and the proposition it expresses are permanently false. To see that (4) really is the corresponding *de dicto* date-sentence, note that the translation of a 1 A.D. utterance of "the present year" is "1 A.D.," used attributively to express the sense of "whatever year includes Christ's birth." "1 A.D." does not refer directly to the year that in fact includes Christ's birth; rather, it refers indirectly to whatever year is the one that has the property of including Christ's birth. Item (4) implies that there is some possible world in which whatever year has the property of including Christ's birth also has the property of not including his birth. This of course is self-contradictory and prevents the proposition expressed by (4) from being equivalent, even materially, to the one expressed by T_3.

This nonequivalence extends to the counterfactual propositions about Christ's birth that are expressible at any time by A-sentence-tokens. A 1986 utterance of

(5) Christ might not have been born earlier than now

expresses a true proposition; but its corresponding *de dicto* date-sentence,

(6) Christ might not have been born earlier than whatever year is 1,985 years later than Christ's birth

is self-contradictory. Of course, if we insert "actually" between "is" and "1,985," the proposition becomes true; but the standard *de dicto* date-theory does not regard the propositional definite descriptions of dates to be world-indexed. Even if we do world-index the descriptions, the *de dicto* date-theory will encounter the same problem as does the *de re* date-theory. This problem arises if we assume the reductionist theory of time. Given this theory, the *de re* translation of the token T_3 of (3) is

(7) Christ might not have been born during 1 A.D.,

where "1 A.D." directly refers to the year-long set of events that includes Christ's birth. (7) does not make the self-contradictory assertion that Christ might not have been born during whatever year-long set of events has the property of including his birth, but it makes a different self-contradictory assertion. It implies

(8) There is some year-long set of events that actually includes Christ's birth but in some merely possible world does not include his birth.

Item (8) violates the principle that sets have their members essentially and therefore the sentence that implies it, (7), is necessarily false.

But it is not merely tokens of sentences such as (3) that express propositions with different truth values than the propositions expressed by their corresponding *de re* date-sentences. Any counterfactual A-sentence-token that asserts that some event might have or might not have occurred now, today, tomorrow, two minutes ago, or the like expresses a proposition with a different truth value than its corresponding *de re* date-proposition. The propositions expressed by July 23, 1986 tokens of

(9) The thunderstorm might not have occurred today

and

(10) There might have been sunny skies today

are both true; but the propositions expressed by their alleged *de re* translations are both false:

(11) The thunderstorm might not have occurred on July 23, 1986.

(12) There might have been sunny skies on July 23, 1986.

Item (11) implies that the set of events that is the direct referent of "July 23, 1986" actually includes the thunderstorm but does not include it in some merely possible world; and (12) implies that the set of events that is the direct referent of "July 23,

1986" does not actually include sunny skies but does include them in some merely possible world. Both assertions are false, since both violate the principle that sets have their members essentially.

If we assume the substantival theory of times, then the *de re* translations of the tokens of (3), (9), and (10) do not violate this principle; for the *de re* translation of, for example, the token T_3 of (3) then entails

(13) There is some possible world in which the moment M that actually includes Christ's birth does not include his birth,

which is self-consistent because M is accidentally occupied by Christ's birth. Thus, the *de re* date-theory and the modally indexed *de dicto* date-theory are logically incompatible with the reductionist theory of time, which will be disturbing to those date-theorists who hold both theories.

But even if we assume the substantival theory of times, the *de re* and modally indexed *de dicto* date-theories are logically incompatible with the truth of some other modal A-sentence-tokens. It is true at any time that

(14) It is necessarily true that some time is present if time exists.

However, the tenseless date-sentences that allegedly express the same propositions as the successive tokens of (14) are one and all false. The alleged translation of the July 23, 1986 utterance of (14) is

(15) It is necessarily true that some time is July 23, 1986 if time exists,

which is false regardless of whether "July 23, 1986" be taken as directly referential or indirectly referential and regardless of whether it be taken as referring to a moment or set of events. Suppose, for example, that it refers directly to the moment M_1. It is then false; for in some possible world W time exists but begins at the moment M_2 that is actually located on July 24, 1986 and in some other world time exists but ends as the moment M_0 that is actually located at July 22, 1986.

Another semantic nonequivalence concerns modal sentences such as

(16) If Friday, April 14, 1989, were not present, it would be past or future if actual at all.

If (16) is uttered on Friday, April 14, 1989, it expresses a true proposition. But the corresponding tenseless date-sentence expresses a necessarily false proposition:

(17) If Friday, April 14, 1989 were not at Friday, April 14, 1989, it would be earlier or later than Friday, April 14, 1989, if actual at all.

Item (17) is necessarily false since there is no possible world in which Friday, April 14, 1989 is not simultaneous with itself. This can be argued more exactly by noting that different uses of the date-description in (16) provide refutations respectively of the *de dicto* and *de re* date-theory. If "Friday, April 14, 1989" is used attributively, so that it is also so used in its first occurrence in the allegedly translating sentence (17), then the *de dicto* date-theory is refuted, since it maintains that the second occurrence of this expression in (17), which translates "is present," is used attributively, which makes (17) necessarily false. But if "Friday, April 14, 1989" in (16) is used referentially, so that it is also so used in its first occurrence in the allegedly translating sentence (17), then the *de re* date-theory is refuted, since it holds that the second occurrence of this expression in (17) is used referentially, which makes (17) necessarily false.

I conclude, then, that modal A-sentence-tokens such as these are so far from expressing propositions logically equivalent to those expressed by the corresponding tenseless date-sentences that they do not even express materially equivalent propositions. But the tenseless date-theory has still more problems.

2.5 The Logical Identities of A-Sentence-Tokens and Tenseless Date-Sentences

A necessary and sufficient condition of one sentence or sentence-token expressing the same proposition as another is that they are logically identical, that is, that they refer in the same ways to the same items and ascribe the same n-adic properties to these items. My argument in this section is that A-sentence-tokens and their corresponding tenseless date-sentences are logically different if only for the reason that they either (1) refer to different temporal positions, (2) refer in different ways to the same temporal positions, or (3) ascribe different temporal n-adic properties. Since the translation condition of logical identity is the finest grained of the four conditions I use (it is finer grained than logical equivalence, sameness of confirmation conditions, and sameness of truth conditions), it is suitable that I here examine the sorts of linguistic constructions that reveal the finest-grained differences in linguistic meaning, oratio obliqua constructions (e.g., "She believes that x is F"). I shall use these constructions to argue against the theory that A-utterances express tenseless beliefs *de dicto* about dates, the theory that they express tenseless beliefs *de re*, about dates, and the theory (of Castañeda) that they express tenseless beliefs about dates that are neither *de dicto* nor *de re*.

According to Bertrand Russell (at one period of his thinking), A-sentence-tokens express *de dicto* date-propositions and convey our *de dicto* beliefs about dates. For example, "When we are told 'Mrs. Brown is not at home', we know the time at which this is said, and therefore we know what is meant. But in order to express explicitly the whole of what is meant, it is necessary to add the date."[53] But this seems plainly false. I can perfectly well understand a token of "Mrs. Brown is not at home" without knowing whether this token occurs at 2:55 P.M. or 3:20 P.M., on June 15 or June 16. I could even—in case of amnesia—fail to know what century it is and still understand what is meant. This is indicated behaviorally by the fact that if I could not answer the question "At what date did the token occur?," I could still respond appropriately, for example, by postponing my trip to Mrs. Brown's house and affirming correctly that "Mrs. Brown is not at home simultaneously with that utterance of 'Mrs. Brown is not at home'."

Our normal ignorance of dates is encapsulated in the nonintersubstitutivity of A-words and date-descriptions in *de dicto* belief reports. A 2:55 P.M., June 15 utterance of

(0) John believes that Mrs. Brown is not at home

does not retain the same truth value if a date-description is substituted for the A-copula:

(1) John believes that Mrs. Brown (is) not at home at 2:55 P.M. on June 15.

for John may well believe at 2:55 P.M. on June 15 that Mrs. Brown is not at home without believing, or even while disbelieving, that she is not at home at 2:55 P.M. on June 15.

The fact that we normally believe what is expressed by A-sentence-tokens without believing what is expressible by the corresponding tenseless date-sentences entails that our A-words do not express tenseless propositional definite descriptions of dates. A second difference in *de dicto* beliefs concerns the token-reflexive B-relations that the A-sentence-tokens possess to the events they are about. A sincere utterance of "Mrs. Brown is not at home" by John entails

(2) John believes that Mrs. Brown (is) not at home simultaneously with his utterance,

whereas John's sincere utterance (at any time) of "Mrs. Brown (is) not at home at 2:55 P.M. on June 15" does not entail (2), which shows that A-sentence-tokens convey a specific sort of temporal information and express a specific sort of temporal beliefs not conveyed and expressed by tenseless date-sentence-tokens.

The direct reference date-theory is able to defend itself against some of these criticisms, since it holds that the date-beliefs expressed by A-sentence-tokens are *de re*, rather than *de dicto*, which is consistent with the fact that we are normally ignorant of the dates of our tokens. The *de re* theory, then, is not so easy to refute as the *de dicto* date-theory. Patrick Grim believes that the *de re* date-theory is so hard to refute that we need to resort to examples involving time machines or beliefs in time machines to construct arguments against it,[54] but I believe there are perfectly natural examples that suffice. Consider a May 7, 1982 utterance of

(3) John believes that the battle is starting today.

Its alleged *de re* date-translation is

(4) John believes of May 7, 1982 that the battle is starting then.

But a May 7, 1982 token of (3) and (4) are not intersubstitutable *salva veritate;* for on May 7, John may believe of May 7 that the battle is starting then without believing of May 7 that it is today. John may mistakenly believe of May 7 that it is long past. Suppose there is a live broadcast portraying the battle on May 7, the day the battle starts. John believes *de re* of the day that is shown or pictured in the broadcast that it is the day the battle starts. But John does not realize that the broadcast is live and that the day shown in the broadcast is *today*. Indeed, John believes and sincerely asserts that the battle is *not* starting today. Item (4) is true, but the May 7 utterance of (3) is false. Moreover, the May 7 utterance of (3) may be true, and (4), false. John may sincerely assert on May 7, "The battle is starting today" and retain this belief while being shown footage of actions taking place on May 7 and while believing of the day shown in the footage that the battle is not starting then. John may be under the misapprehension that the footage is one day old.

The fact that the *de re* date-descriptions are not intersubstitutable in belief-reports with tokens of "today" shows that they have different semantic contents. Being related by the relation of believing to the direct referent of "May 7, 1982" is not being related to the same thing that one is related to when one is related by this relation to what is expressed by the May 7, 1982 token of "today." The token of "today" expresses something different from, or additional to, what is expressed by the referential use of "May 7, 1982."

But this argument would not refute the direct reference date-theorists who hold that A-indexicals do not express two-termed beliefs *de re* but three-termed beliefs, whose third term is the cognitive significance of the A-indexical. However, I have already critically discussed David Kaplan's and Nathan Salmon's theories of this three-termed belief relation (section 2.3) and shall not repeat these criticisms here.

Before I draw this chapter to a conclusion and proceed to the token-reflexive theory, I shall consider one more theory, the date theory developed by Castañeda in his 1960s articles.[55] Castañeda's theory is distinctive in that he claims that uses of temporal indexicals directly refer to dates but that the beliefs about dates they express are neither *de re* nor *de dicto* but of a third sort, one that is uniquely captured in oratio obliqua by the quasi indicator "then." The occurrence of "then" in the following sentence attributes to the person mentioned "a purely indexical reference to the time of his thinking":

(5) On May 15, 1911, the German emperor believed that it was raining then.

"Clearly," Castañeda continues, "the German Emperor may have had no idea at all as to the date on which it was raining then, and he may have also failed to have any other non-indexical description that uniquely characterized May 15, 1911."[56] But this is not to say that his belief is *de re,* for there is another sort of belief. Castañeda's argument for this third sort of belief is only implicitly developed for temporal beliefs, and his main argument is for the irreducibility of first-person beliefs to beliefs *de dicto* or *de re* about oneself. The argument for irreducible temporal-indexical beliefs can be developed if we construct an argument of the same form as Castañeda's argument for irreducible *de se* beliefs. Castañeda points to alleged entailment relations between certain sentences; he avers that a sentence that attributes belief *de se,* such as

(6) The editor of *Soul* believes that he himself is a millionaire

entails a sentence such as

(7) The editor of *Soul* believes what he would express by saying, "I am a millionaire"

and therefore is semantically irreducible to a sentence that attributes a belief *de dicto,* such as

(8) The editor of *Soul* believes that whoever is the editor of *Soul* is a millionaire

and irreducible to a sentence that attributes a belief *de re,* such as

(9) The editor of *Soul* believes of the editor of *Soul* that he is a millionaire,

since neither (8) nor (9) entail (7). The analogous argument for irreducible temporal-indexical beliefs would be that sentence (5), about the emperor, attributes such a belief to the emperor and entails

(10) On May 15, 1911, the German emperor believed what he would then express by saying, "It is raining now"

and that the two sentences that ascribe *de dicto* and *de re* belief to the emperor, namely,

(11) On May 15, 1911, the German emperor believed that it was raining on whatever day is May 15, 1911

and

(12) On May 15, 1911, the German emperor believed of May 15, 1911 that it was rainy,

do not entail (10).

In (5) and (6), "then" and "he himself" are *quasi indicators,* which have the following properties, among others: (i) they do not express indexical references made by the speaker, (ii) they require an antecedent to which they refer back (as "then" refers back to "May 15, 1911") but are not replaceable by their antecedents, (iii) they are not intersubstitutable with names or descriptions, and (iv) they "represent uses" of indexicals by somebody p such that they are means of making p's "indexical reference both interpersonal and enduring, yet preserving it intact."[57]

If this theory is correct, then the tenseless date-theory is vindicated; for the failure of reports of beliefs *de dicto* and *de re* about dates to capture what we believe when we use temporal indexicals would then be consistent with the tenseless date-theory, for date-theorists could take refuge in the claim that our beliefs about dates are of a unique temporal-indexical sort and *for this reason* are not captured by reports of beliefs *de dicto* and *de re*. But I believe that Castañeda's theory cannot save the date-theorists, since his argument that the quasi indicator "then" (which refers to dates) preserves intact the indexical reference of uses of "now" is unsound. I shall demonstrate this in terms of the example Castañeda has treated most fully, that involving "the height of the Chrysler Building," discussed in his "Omniscience and Indexical Reference."[58]

At t_1, the height of the Chrysler Building is 1,046 feet; and at t_2 it is extended, by the addition of an antenna, to 1,086 feet:

> The question . . . is whether or not a person can know at time t_1 (prior to the extension of the antenna) a proposition that he would express at t_2 (after the extension) by uttering a sentence containing an indicator, e.g., "Now the Chrysler Building is 1,086 feet tall." Once again, the answer is "yes," and a way of finding one formulation of that proposition is the method illustrated above, in which we employed principle (P). Thus, suppose that
>
>> (13) Kretzmann knows at t_1 that: the Chrysler Building is 1,046 feet high at t_1, and at t_2 it will have a 40-foot antenna extended from its tip, and that the man who makes the extension knows at t_2 that the Chrysler Building is 1,086 feet high then.
>
> Clearly, if (13) obtains, Kretzmann knows of the change in height without having to change his knowledge.[59]

Principle (P) is

> (P) If a sentence of the form "X knows that a person Y knows that . . ." formulates a true statement [proposition], then the person X knows the statement formulated by the clause filling the blank " . . ."

If (13) and (P) are both true, does Kretzmann know at t_1 the proposition that the man who makes the extension (call him M) would express at t_2 by uttering, "Now the Chrysler Building is 1,086 feet tall"? No, for what Kretzmann relevantly knows at t_1, *that the Chrysler Building is 1,046 feet at t_1 and that the man who makes the extension knows at t_2 that the Chrysler Building is 1,086 feet high then,* does not entail

> (14) The Chrysler Building is not 1,046 feet tall at the present time,

whereas what M knows by knowing the proposition expressed at t_2 by "Now the Chrysler Building is 1,086 feet tall" does entail (14). Consequently, the occurrence of "then" in (13) does *not* "preserve intact" the semantic content or reference of the occurrence of "now" in M's utterance of "Now the Chrysler Building is 1,086 feet tall" at t_2. Accordingly, Castañeda's theory of indicators and quasi indicators cannot be used to vindicate the tenseless date-theory of A-sentences.

Patrick Grim would object that my account misrepresents Castañeda's theory. According to Grim, Castañeda does not hold "that *what* is expressed at one time using an indexical can *itself* be expressed at another time using a quasi-indicator in *oratio recta*."[60] This is true enough; but Grim erroneously concludes from this that

"Castañeda holds that the *fact* that someone knows something that they might express using an indexical is a *fact* that can be expressed by others or at another time using a quasi-indicator in *oratio obliqua*."[61] But Castañeda does not hold this but instead, that *what* is expressed at one time using an indexical can *itself* be expressed at another time using a quasi indicator in oratio obliqua. This is precisely what is entailed by Castañeda's statement that "a person can know at time t_1 . . . a proposition that he would express at t_2 . . . by uttering a sentence containing an indicator [i.e., 'now']. . . . A way of finding one formulation of that proposition [is found in the oratio obliqua construction (13)]." If Grim's interpretation were correct, this quotation would have read instead, "a person can know at time t_1, not the *proposition* that he would express at t_2 by uttering a sentence containing 'now', but merely *the fact that he knows at t_2 a proposition* that he would express at t_2 by uttering a sentence containing 'now'." . . . and one formulation of *the fact* that he knows at t_2 that proposition is found in the oratio obliqua construction (13)." I think, accordingly, that Grim gives us no reason to think my argument against Castañeda to be unsound. In chapter 4, I shall develop a theory of quasi indicators that is consistent with the tensed theory of time.

2.6 A Similarity Between A-Sentence-Tokens and Tenseless Date-Sentences: Intersubstitutivity in Extensional Contexts

The differences between A-sentence-tokens and their corresponding tenseless date-sentences discussed in the preceding four sections suffice to show that these corresponding items express different propositions. This is the case even though these corresponding tokens and sentences *meet* some further necessary conditions for expressing the same proposition, most notably the condition of intersubstitutability *salva veritate* in extensional contexts. This condition may be applied to present tensed A-sentences and tenseless date-sentences as follows:

> EI: Each present tensed A-sentence qua tokened in an extensional context at some date D has the same truth value as a tenseless date-sentence that is formed by replacing the present tensed aspect of the A-sentence by an expression that refers to the date D and ascribes a relation of simultaneity between D and the event to which the A-token refers.

For example, if "John is now walking" is tokened in an extensional context at noon, E.S.T., June 8, 1990, then a sentence with the same truth value as this token can be formed by replacing the "now" and the present tense of the copula with the expression "at noon, E.S.T., June 8, 1990." The date-sentence formed will be "John (is) walking at noon, E.S.T., June 8, 1990." For each extensionally occurring token of this A-sentence that occurs at a different date, there will be a different date-sentence that is intersubstitutable with it *salva veritate*.

A converse relation also holds. Each extensionally occurring date-sentence that ascribes a simultaneity relation between some event E and some date D has the same truth value as would a tokening of the corresponding present tensed A-sentence on the date D. The corresponding A-sentence is formed by eliminating the date-expression in the date-sentence, adding a present tense to the copula and/or verb, and perhaps also adding an A-word like "now."

It is important to note that this intersubstitutivity relation only holds between tenseless date-sentences and A-sentences in extensional contexts. Truth values are not preserved by substitutions in intensional contexts; this was shown in the last two sections for sentences prefixed by modal operators (2.4) and by epistemic operators (2.5).

Intersubstitutivity *salva veritate* in extensional contexts is the principal semantic similarity between A-sentence-tokens and tenseless date-sentences, and explains why so many philosophers have been led to the view that A-sentence-tokens express the same propositions as date-sentences and/or are translatable, analyzable, or paraphrasable by them. However, this similarity is a necessary but *insufficient* condition for these linguistic formations to express the same propositions; and the date-theorists have failed to study completely or adequately the other tenseless semantic rules of these formations, which reveal they cannot express the same propositions. The differences in truth conditions, confirmation conditions, entailment relations, and logical identities discussed in the foregoing sections make it apparent that different propositions are involved.

The fact that the semantic differences between A-sentence-tokens and their corresponding tenseless date-sentences far outweighs their semantic similarities may be highlighted by showing that the differences prevent these formations from sustaining even the very *weakest* of the translation–analysis–paraphrase relations postulated by the date-theorists, namely, the paraphrase relations postulated by Quine. Quine does not wish to claim that the date-sentences are "synonymous" with A-sentence-tokens, that they constitute "uniquely right analyses" of them, or even that they are logically equivalent. Instead, they share a similar function: "There opens the question of how this eternal sentence is related to the given utterance of the non-eternal one. . . . The eternal sentence will be one that the original speaker could have uttered in place of his original utterance in those original circumstances without detriment, so far as he could forsee, to the project he was bent on."[62] The relevant project upon which I am bent in uttering a (declarative) A-sentence is to communicate some temporal information about an event. Quine believes that if I had uttered the corresponding B-sentence instead, I would have explicitly communicated the temporal information implicitly conveyed by my original utterance and thereby would accomplish the same project, only this time in an explicit manner. But this is a mistake. If I uttered the date-sentence instead, I would have conveyed different temporal information and would have accomplished a different project. By uttering at 3:37 P.M., E.S.T., August 2, 1985, "The car is exploding," I aim to convey the information that the car is exploding simultaneously with my utterance and therefore (by implication) that my addressees should run for their lives. But if I had instead uttered the date-sentence intersubstitutable with it *salva veritate,* the sentence "The car (is) exploding at 3:37 P.M., E.S.T., August 2, 1985," I would have failed to accomplish this project. For this sentence does not convey the information that the car is exploding *when I am saying this* and therefore fails to convey the implied warning that my addressees should immediately run away. The project of communicating that the car is exploding *simultaneously with my utterance* is a very different project than communicating that it is exploding *1,984 years, 7 months, 1 day, 15 hours, and 37 minutes after Christ's birth.*

A second reason for thinking that Quine's paraphrase relations do not hold is that the paraphrases are meant to be "aids to understanding the referential work of language." Now Quine believes that A-sentence-tokens tacitly refer to dates and hence that their date-sentence paraphrases aid our understanding of this reference by making it explicit. However, as we have seen in previous sections, this is false; tokens of sentences such as "The car is exploding" or "Henry is ill" do not refer to dates even tacitly. Consequently, paraphrasing them by sentences that do refer to dates does not aid but *hinders* our understanding of the referential work of language. Instead of making explicit the original's reference, the date-sentences introduce a reference not present in the original and omit a reference or a predication present in the original (i.e., the predication of a simultaneity relation between the event and the utterance). I think these considerations make it plain that the extensional inter-substitutivity *salva veritate* of A-sentence-tokens and their corresponding date-sentences does not even suffice to allow Quine's paraphrases to go through.

The arguments in this chapter are one step on the road to establishing that A-sentence-tokens refer to, or predicate, A-positions. The arguments that they are untranslatable by tenseless date-sentences have aimed to show that the A-tokens do not express tenseless date-propositions and therefore that one of the two basic detenser theories of A-sentences is false. The second basic detenser theory, the token-reflexive theory, holds that A-sentences predicate merely *B-relations* of events and *their own utterances*. None of the arguments presented in this chapter suffices to refute the token-reflexive theory. In fact, it is conceivable that a defender of the token-reflexive theory could agree with almost everything in this chapter. Accordingly, my next task is to develop a second set of arguments that apply uniquely to the token-reflexive theory. Once this is done, the positive argument that A-sentences relate to A-positions can be constructed (chapter 4). This shall complete part I. In part II, I shall develop the argument for presentism, that is, the argument that even tenseless sentences ascribe a presentness-involving property and that the temporal properties ascribed constitute an absolute (and not a relativistic) time series.

Notes

1. The quotations are from Milton Fisk, "A Pragmatic Account of Tenses," *American Philosophical Quarterly* 8(1971): 94; Paul Fitzgerald, "Nowness and the Understanding of Time," Boston Studies in the Philosophy of Science, no. 20 (Dordrecht: Reidel, 1974), 268; Nelson Goodman, *The Structure of Appearance* (Cambridge, MA: Harvard University Press, 1951), 296; David Kaplan, "On the Logic of Demonstratives," in *Contemporary Perspectives in the Philosophy of Language,* ed. P. A. French et al. (Minneapolis: University of Minnesota Press, 1977), 403; John Perry, "Frege on Demonstratives," *Philosophical Review* 86(1977): 494; W.V.O. Quine, *Elementary Logic* (Cambridge, MA: Harvard University Press, 1980), 6; Bertrand Russell, Review of *Symbolic Logic and Its Application* by Hugh MacColl, *Mind*

15(1906): 256–57; Clifford Williams, "'Now', Extensional Interchangeability, and the Passage of Time," *Philosophical Forum* 5(1974): 405–6.

2. Gottleib Frege, "The Thought," in *Essays on Frege,* ed. E. D. Klempke (Urbana: University of Illinois Press, 1968), 516–17.

3. Ibid., 516.

4. John Mackie, "Three Steps Towards Absolutism," in *Space, Time, and Causality,* ed. Richard Swinburne (Dordrecht: Reidel, 1981).

5. W. H. Newton-Smith, *The Structure of Time* (London: Routledge & Kegan Paul, 1980), 6–7.

6. Bertrand Russell, *Human Knowledge* (New York: Simon and Schuster, 1948), 271.

7. Hans Reichenbach, *The Philosophy of Space and Time* (New York: Dover, 1950); Adolf Grünbaum, *Philosophical Problems of Space and Time* (Dordrecht: Reidel, 1973); Mario Bunge, *Foundations of Physics* (New York: Springer-Verlag, 1967); Ian Hinckfuss, *The Existence of Space and Time* (Oxford: Clarendon, 1975).

8. Newton-Smith, *Structure of Time,* 9–10.

9. Isaac Newton, *Newton's Philosophy of Nature,* ed. H. S. Thayer (New York: Hafner, 1953); Richard Swinburne, *Space and Time,* 2d ed. (New York: St. Martin's, 1981); Sydney Shoemaker, "Time Without Change," *Journal of Philosophy* 66(1969): 363–81; Quentin Smith, *The Felt Meanings of the World: A Metaphysics of Feeling* (West Lafayette, Ind.: Purdue University Press, 1986).

10. A. N. Prior, *Papers on Time and Tense* (Oxford: Oxford University Press, 1968).

11. David Lewis, *On the Plurality of Worlds* (Oxford: Basil Blackwell, 1986), 83.

12. Perry, "Frege on Demonstratives," 493.

13. Ibid.

14. The direct reference theory of spatial indexicals, such as "here," must also be modified to accommodate the differences between pronomial and adverbial uses. I argue this in my "Temporal Indexicals," *Erkenntnis* 32(1990): 5–25.

15. W.V.O. Quine, *Word and Object* (Cambridge: Massachusetts Institute of Technology Press, 1960), 208.

16. Quine, *Elementary Logic,* 92.

17. Ibid.

18. Cf. Richard Taylor, *Metaphysics* (Englewood Cliffs, N.J.: Prentice-Hall, 1963); and Richard Gale, *The Language of Time* (New York: Humanities, 1968).

19. Williams, "'Now', Extensional Interchangeability, and the Passage of Time."

20. J.J.C. Smart, "Time and Becoming," in *Time and Cause,* ed. P. van Inwagen (Boston: Reidel, 1980), 5.

21. J.J.C. Smart, private communication, January 5, 1988.

22. Ibid.

23. D. H. Mellor, *Real Time* (Cambridge: Cambridge University Press, 1981), 76.

24. See H. Field, "Mental Representations," *Erkenntnis* 13(1978): 9–61; M. Platts, *Ways of Meaning* (London: Routledge & Kegan Paul, 1979); J. McDowell, "Truth Conditions, Bivalence, and Verification," *Truth and Meaning,* eds. G. Evans and J. McDowell (Oxford: Oxford University Press, 1976); D. Davidson, "Truth and Meaning," *Synthese* 17(1967): 304–23; idem, "Belief and the Basis of Meaning," *Synthese* 27(1974): 309–24.

25. Nicholas Asher, "Trouble with Extensional Semantics," *Philosophical Studies* 47(1985): 1–14.

26. Stephen Schiffer, *Remnants of Meaning* (Cambridge: Massachusetts Institute of Technology Press), chap. 5.

27. See my "Problems with the New Tenseless Theory of Time," *Philosophical Studies*

52(1987): 371–92. Oaklander's objection is presented in his "Defense of the New Tenseless Theory of Time," *Philosophical Quarterly* 41(1991): 26–38.

28. Oaklander, "Defense of the New Tenseless Theory of Time," (unpublished version) 22. My italics.

29. Ibid., 23.

30. Adolf Grünbaum, *Modern Science and Zeno's Paradoxes* (Middletown, CT: Wesleyan University Press), 20.

31. William Lycan, "Eternal Sentences Again," *Philosophical Studies* 26(1974): 412.

32. Ibid., 418.

33. See Keith Donellan, "Reference and Definite Descriptions," in *The Philosophy of Language,* ed. A. P. Martinich (New York: Oxford University Press, 1985)

34. Kaplan, "On the Logic of Demonstratives," p. 60.

35. Hector-Neri Castañeda, "Direct Reference, Realism, and Guise Theory," 1984 mimeograph, pp. 28–28; Palle Yourgrau, "Frege, Perry, and Demonstratives," *Canadian Journal of Philosophy* 12(1982): 728.

36. See Nathan Salmon, *Frege's Puzzle* (Cambridge Massachusetts Institute of Technology Press, 1986); Howard Wettstein, "Has Semantics Rested on a Mistake?," *Journal of Philosophy* 83(1986): 185–209; Michael Tye, "The Puzzle of Hesperus and Phosphorus," *Australasian Journal of Philosophy* 56(1978): 219–24; Raymond Bradley and Norman Swatz, *Possible Worlds* (Indianapolis: Hackett, 1979), 191–92; Tom McKay, "On Proper Names in Belief Ascriptions," *Philosophical Studies* 39(1981): 287–303; R. M. Sainsbury, "On a Fregean Argument for the Distinctness of Sense and Reference," *Analysis* 43(1983): 12–14.

37. Salmon, *Frege's Puzzle,* 24–40.

38. Ibid., p. 111.

39. Ibid.,

40. Kaplan, "Demonstratives."

41. Ernest Sosa, "Consciousness of the Self and of the Present," in *Agent, Language, and the Structure of the World,* ed. James E. Tomberlin (Indianapolis: Hackett, 1983), 131–45.

42. Hector-Neri Castañeda, *On Philosophical Method* (Bloomington, IN: Noûs Publications, 1980).

43. Michele Beer, "Temporical Indexicals and the Passage of Time," *Philosophical Quarterly* 38(1988): 162.

44. Ibid.

45. Michelle Beer and Richard Gale, "An Identity Theory of the A- and B-Series," *Dialectics and Humanism* (forthcoming).

46. Oaklander, "A Defense of the New Tenseless Theory of Time."

47. Beer, "Temporal Indexicals," 162.

48. Ibid.

49. Cf. my "The Co-reporting Theory of Tensed and Tenseless Sentences," *Philosophical Quarterly* 40(1990): 223–32.

50. I have substituted "token" for Goodman's "sentence," for Goodman uses "sentence" to mean sentence-token. See Goodman, *Structure of Appearance,* 287ff.

51. Ibid., 296–97.

52. Steven Boer and William Lycan, "Who, Me?," *Philosophical Review* 89(1980): 433–34.

53. Russell, Review of *Symbolic Logic and Its Application* by Hugh Maccoll.

54. Patrick Grim, "Against Omniscience: The Case from Essential Indexicals," *Noûs* 19(1985): 157–59.

55. For references, see James E. Tomberlin, ed., *Agent, Language, and the Structure of the World* (Indianapolis: Hackett, 1983), 467–76.

56. Hector-Neri Castañeda, "Indicators and Quasi-Indicators," *American Philosophical Quarterly* 4(1967): 96. I omit the asterisk after "then," which Castañeda uses to show that "then" occurs as a quasi indicator.

57. Ibid., 85.

58. Idem, "Omniscience and Indexical Reference," *Journal of Philosophy* 64(1967): 203–9.

59. Ibid., 208–9.

60. Patrick Grim, "Against Omniscience," 164.

61. Ibid.

62. Quine, *Word and Object*, 208.

3

The Untranslatability of A-Sentences by Tenseless Token-Reflexive Sentences

3.1 The Old and New Token-Reflexive Theories of A-Sentences

In this section I first explain the old and new token-reflexive theories of A-sentences and then show that the new theory reduces, on pain of self-contradiction, to the old theory. This reduction will simplify my argument in the remaining sections of this chapter, which will need be directed only against the old token-reflexive theory.

The old token-reflexive theory of A-sentences is usually regarded as originating in 1947 with Hans Reichenbach's *Elements of Symbolic Logic*. Reichenbach there argued that "'now' means the same as 'the time at which this token is uttered'."[1] However, the basic idea behind the token-reflexive theory had appeared as early as 1936 with William Kneale's "Is Existence a Predicate?" Kneale claimed that each token of a demonstrative like "now" denotes a thing and

> denotes whatever it does denote by locating that thing as in a certain relation to itself which anything denoted by a token of that type must have to the token denoting it. If anyone else understands my use of a demonstrative word, he does so by apprehending the particular sound which I utter and considering what thing can be in a certain relation to that particular sound. The rules for the use of tenses in verbs must be explained in a similar fashion.[2]

Michael Dummett, B. Mayo, Marc Temin, D. C. Williams, and others have espoused versions of the old token-reflexive theory; but the most completely worked-out version of this old theory appears in J.J.C. Smart's "River of Time" and *Philosophy and Scientific Realism*.[3] In the latter work, Smart writes:

> Let us replace the words "is past" by the words "*is* earlier than this utterance." (Note the transition to the tenseless "is.") Similarly, let us replace "is present" and "now" by "*is* simultaneous with this utterance," and "is future" by "*is* later than this utterance." By "utterance" here, I mean, in the case of spoken utterances the actual sounds that are uttered. In the case of written sentences (which extend through time) I mean the earliest time slices of such sentences (ink marks on paper). Notice that I am here talking of self-referential *utterances*, not self-referential *sentences*.

(The same sentence can be uttered on many occasions.) We can, following Reichenbach, call the utterance itself a "token," and this sort of reflexivity "token-reflexivity." Tenses can also be eliminated, since such a sentence as "he will run" can be replaced by "he *runs* at some future time" (with tenseless "runs") and hence by "he *runs* later than this utterance." Similarly, "he runs" means "he *runs* (tenseless) simultaneous with this utterance," and "he ran" means "he *runs* (tenseless) earlier than this utterance." All the jobs which can be done by tenses can be done by means of the tenseless way of talking and the self-referential utterance "this utterance."[4]

On this theory, some tenseless sentence S_1 translates some tensed sentence S_2; it is not the case that a different tenseless sentence translates each different token of the tensed sentence S_2. This is the thrust of Smart's remark that "such a sentence as 'he will run' can be replaced by . . . 'he *runs* [tenseless] later than this utterance'." This suggests that each token of "he will run," not just one of its tokens, is translatable by a token of "he (runs) later than this utterance."

Smart and the other proponents of the old token-reflexive theory do not appear to recognize that two different versions of this theory have been advanced. Some of the semantic implications of these two versions can be made explicit in the following ways.

The version suggested by Smart's remarks involves the idea that A-word-tokens refer to themselves and predicate B-relations. The self-reference of the A-tokens may be conceived in one of two ways, depending upon one's theory of reference. One may represent them as referring to themselves directly, without the intermediary of a sense or propositional element, or as referring to themselves indirectly, by expressing the sense *this token*. Moreover, one may view them as referring only to themselves or as referring to the whole sentence-token of which they are a part. The predicative function of the A-word-tokens involves the dyadic properties of simultaneous, earlier, and later. The event referred to by the sentence-token is related by one of these properties to the sentence-token itself (or to the A-word-token that is a part of the sentence-token), such that the event is asserted to be simultaneous with (or earlier or later than) the token.

A second version of the old token-reflexive theory is suggested by the remarks of William Kneale. He claims that each token of an A-word refers to a "thing" and locates "that thing as in a certain relation to itself [the token]". This implies that A-tokens refer not only to themselves but also to "things." He does not say what these "things" are; but they are presumably times, moments or sets of simultaneous events. (The "thing" cannot be the event the sentence is about, for this event is already referred to by the rest of the sentence. Surely, the "now" in "John is now running" does not refer to John's running.) This intepretation accords with Reichenbach's remark that "now" means "the time at which this token is uttered," where the referent of "the time" is presumably the moment or set of simultaneous events to which the token belongs. If one is not a nominalist, one will find it plausible to suppose that this account of A-words implies that present tensed A-words express the indexical-containing, propositional, definite description *whatever set of events (or moment) includes this token*. When "now" or another present tensed A-term has a pronomial function, this definite description is all that is expressed; but A-copulae, verbs, adverbs, and predicates must also have a predicative function. For example, "John is asleep now" cannot mean "John (is) asleep whatever set of events includes this token." It must mean "John's (being) asleep (is) included in whatever set of simultaneous events includes this token."

The first and most completely developed version of the *new* token-reflexive theory of A-sentences appeared in D. H. Mellor's 1981 book, *Real Time*. Mellor shares with the old token-reflexive theorists the idea that "the tenseless facts that fix the truth-values of tokens of simple tensed sentences and judgments . . . include facts about the tokens themselves—their relative whereabouts in the B series"; and, "this account of what makes them true is called a 'token-reflexive' account".[5] But Mellor denies the translatability thesis. Mellor is concerned with showing that A-sentences and their tokens are untranslatable by a particular sort of tenseless sentence and token, namely, those that state the truth conditions of the A-sentences and tokens. The truth conditions of any token S of "It is now 1980" are that S is true if and only if it occurs in 1980. But the sentence stating this truth condition, namely, the sentence "S occurs in 1980," is such that neither it nor any of its tokens translates S or the sentence-type "It is now 1980." The reason, Mellor, argues, is that the tenseless sentence and tokens have truth conditions different from those of the tensed ones. S is true if and only if it occurs in 1980 (and "It is now 1980" is true if and only if it is tokened in 1980); but these are not the truth conditions of "S occurs in 1980" or any of its tokens. Rather, the tenseless sentence or token is true regardless of when it is tokened, in 1980 or 1990, and requires merely that S occur in 1980.

It seems to me this theory is self-contradictory and that the removal of one its contradictory theses results in its reduction to the old token-reflexive theory of A-sentences. Let us consider S and any token T of "S occurs in 1980." According to Mellor, T does not translate S, since S and T have different truth conditions. But this contradicts Mellor's account of S and T, since Mellor holds that S is true if and only if *S occurs in 1980* and that T is true if and only if *S occurs in 1980.*[6] They have the same truth conditions! Mellor was led to this self-contradictory position through fallaciously inferring from

(1) It is true of S but not of T that *it is true if and only if it occurs in 1980*

to

(2) S and T have different truth conditions.

This is mistaken, since (1) does not state a difference in the truth conditions of S and T but a difference in the relation of their common truth condition, namely, *that S occurs in 1980,* to S and T. Item (1) states that these truth conditions *are about* S and *are not about* T.

Mellor responded to an earlier statement of this criticism by saying that it is wrongheaded to talk of truth conditions being "about" sentence-tokens or other phenomena.[7] But the language of "aboutness" is precisely the language Mellor uses throughout *Real Time* to describe the relation of truth conditions (which Mellor identifies with *facts*) to sentence-tokens and other phenomena. For example, Mellor uses such expressions as "facts *about* them" and also writes, "The tenseless facts that fix the truth-values of tokens of simple tensed sentences and judgments . . . includes facts *about* the tokens themselves."[8]

But suppose we adopt Mellor's later position that facts or truth conditions are not "about" phenomena. This would not remove the contradiction that S and T both do and do not have the same truth conditions or the fallacious inference from (1) to (2). One reason (indeed, the only reason I can think of) for denying that truth conditions are *about* sentence-tokens or other phenomena is that they, instead, *include* these phenomena. (Mellor is not forthcoming on this matter.) But if we substitute "include" for "about" (or some other relevant relation R for "about"), the stated criticism of Mellor remains sound. We would then say that (1) does not state that S and T have different truth conditions but merely that their common truth condition, namely, *that S occurs in 1980, includes* S but not T (or stands in R to S but not to T).

L. Nathan Oaklander responded to my earlier statement of this criticism of Mellor by saying that Mellor holds merely that the relevant sentence-types are untranslatable and allows that their tokens, such as S and T, have the same truth conditions.[9] It follows, according to Oaklander, that "the alleged inconsistency attributed to Mellor vanishes for there is no inconsistency in claiming that tensed and tenseless sentence-types have different truth conditions, while also claiming that tensed and tenseless sentence tokens have the same truth conditions."[10] But Oaklander misunderstands Mellor's theory, which is that the sentence tokens also have different truth conditions. For example, Mellor writes:

> Now the truth conditions of *tokens* of spatially tensed sentence types vary, as we have seen, with their spatial position, and those of temporally tensed types vary with their temporal position. So therefore must the truth conditions of *tokens* of their translations. But what makes sentence types tenseless, as we saw in chapters 1 and 2, is that the truth conditions of their *tokens* do *not* vary in this way . . . . No tenseless sentence, therefore, can have *tokens* whose truth conditions are everywhere and always the same as [the truth conditions of *tokens*] of a tensed sentence, because by definition the latter vary from place to place or time to time and the former do not.[11]

The type/token distinction, consequently, cannot be used to demonstrate the consistency of Mellor's theory. But Oaklander makes a second attempt to show that it is consistent. Oaklander adds that even if the tensed and tenseless tokens have the same truth conditions, it still follows from Mellor's own principles that they do not translate one another, since the tenseless and tensed tokens have *a different usage*.[12] Oaklander alleges that Mellor's theory is that differences in meanings correspond to differences in usage, not just to differences in truth conditions. But this also is a misunderstanding of Mellor's theory, for Mellor defines differences in usage and meaning of the relevant sentence tokens in terms of differences in their truth conditions. In the case of temporally and spatially tensed sentence tokens, "correct usage *is* explained by people knowing how the truth of what they say depends on when and where they say it, in particular the *different* meanings of different sentences are differentiated, as they are not in mathematics, by their different truth conditions."[13] In the case of the logically contingent sentences or sentence-tokens, if they have the same truth conditions, they have the same meaning and usage.

I believe that the contradiction is, indeed, present in Mellor's theory and can be removed only by altering his theory. If we remove from his theory the fallacious inference from (1) to (2) and thereby remove the contradiction that S and T both do and do not have the same truth conditions, then Mellor's theory reduces to a version of the *old* token-reflexive theory of A-sentences, that tokens of these sentences are translated by tokens of tenseless token-reflexive sentences. Since a token S of "It is now 1980" has the same truth conditions as a token T of "S occurs in 1980," S is translated by T (and by any other token of "S occurs in 1980"). But this revised version of Mellor's theory differs from Smart's and Reichenbach's on several scores, one being that it does not license the inference that A-sentence-types (as well as their tokens) are translated by tenseless sentence-types. Since Mellor's candidate tenseless sentence-types mention a certain token of an A-sentence-type, such as the token S, they translate only the token they mention and not other tokens of that A-sentence-type. Further comparisons between the revised Mellor theory and Smart's and Reichenbach's will be made below.

John Searle offers a token-reflexive theory of A-sentences in his 1983 book, *Intentionality,* that at first glance seems to be a *new* version of the theory, one that denies the translatability claim. Searle begins by claiming that A-word-tokens express senses and that the sentence-tokens to which they belong express token-reflexive propositions: "Each proposition expressed [by an utterance of an A-sentence] is self-referential to the utterance in which it is expressed."[14] But Searle disagrees with the claim of the old token-reflexive theorists that A-words mean, and are translated by, the tenseless token-reflexive phrases. For Searle, the latter phrases merely state the truth conditions of the A-words. He avers that "now" is not "synonymous with 'the time of this utterance'. [It] could not be synonymous because the self-referentiality of the original is shown but not stated, and in the statement of the truth conditions we have stated it and not shown it."[15] But Searle then *reintroduces* the old token-reflexive theory (in some form) by showing how tenseless translations can be constructed for the A-words. He indicates that if we want to introduce a synonym or translation of the A-words, an asterisk (*) can be introduced to express, without stating, the fact that the word is self-referential. For Searle, "now" is translated by "at *cotemporal," where "*cotemporal" refers to the time of its utterance and shows, without stating, its self-referentiality. Thus, the A-sentence "I am now hungry" is translated by "I (am) hungry at *cotemporal."

These reflections on Mellor's and Searle's theories, the only two that might seem to be instances of a *new* token-reflexive theory, show that there is no real or self-consistent new token-reflexive theory of A-sentences and therefore that the target of my arguments in this chapter need be only the various versions of the old token-reflexive theory. I have shown that there are at least four versions of the old token-reflexive theory, which may be identified respectively with Smart's, Reichenbach's, Mellor's (revised) and Searle's theories. The differences among them are illustrated by their different translations of a token U_1 of "The tornado is approaching." U_1 is translated respectively by tokens of

(1) The tornado (is) approaching at the moment/set of events at which this token is uttered. (Reichenbach)

(2) The tornado (is) approaching simultaneously with this utterance. (Smart)

(3) The tornado (is) approaching simultaneously with U_1. (Mellor)

(4) The tornado (is) approaching at *cotemporal. (Searle)

The theories of Reichenbach, Smart, and Searle provide for translations of A-sentence-types by tenseless token-reflexive sentence-types; but, as I have mentioned, the revised Mellor theory provides translations only of a given token of an A-sentence-type by (any token of) a tenseless token-reflexive sentence-type. Items (1), (2), and (4)—but not (3)—translate the sentence-type "The tornado is approaching." In this respect, the revised Mellor theory is unique among the three in being similar to the translation thesis of the date-theory of A-sentences discussed in chapter 2.

In the following four sections, I shall show that A-sentence-types and tokens have truth conditions, confirmation conditions, entailment relations, and logical identities different from those of the tenseless sentence-types and tokens illustrated by (1)–(4) and therefore that the latter cannot translate the former. In section 3.2, on truth conditions, I shall discuss sentences about times at which there are no sentence-tokens. In section 3.3, on confirmation conditions, I shall discuss first-person sentences that express propositions stating that the speaker is not saying anything. Sections 3.4 and 3.5, on logical equivalence and logical identity, discuss some differences between A-sentences and token-reflexive sentences in general. The concluding section, 3.6, shows that arguments similar to those used in sections 3.2–5 can be employed to refute the psychological theory of A-sentences developed by Grünbaum and others.

3.2 The Truth Conditions of A-Sentences and Tenseless Token-Reflexive Sentences

The A-sentences I shall study in this section are about past or future truths concerning times when no language users exist. It is important to study these sentences, since many of the A-sentences in the physical sciences (e.g., cosmology, geology, evolutionary biology) are about past or future eras when no language users existed or will exist and entail other A-sentences about past or future truths about these eras. For example, the assertion by Hawking and Ellis that "the universe contains a singularity in the past"[16] is about an event that occurred when no language users were present, and entails such sentences as "It is now true that the universe contains a singularity in the past." Ordinary language also contains such A-sentences; for example, it may be asserted, "It will be true after World War III that cockroaches are alive but humans are not." It is manifest that these sentences are logically coherent and are true under some possible conditions. Consequently, if the token-reflexive sentences that allegedly translate them or their entailments are true under no possible conditions, that will prove good reason for rejecting the token-reflexive account of A-sentences.

Let us isolate for study an A-sentence that is more or less immediately entailed by many scientific and ordinary A-sentences about past times when no language users existed, that is,

(1) It was true that the era devoid of linguistic utterances is present.

This sentence is a past tensed ascription of a truth value *true* to the truth vehicle expressed by the clause following the operator "It was true that." The appropriate interpretation of the nature of this truth vehicle is not indicated by the sentence itself; but we can know that this sentence is true only if the truth vehicle is not a linguistic utterance, for it is logically impossible for there to be a true utterance of "The era devoid of linguistic utterances is present." A necessary truth condition of (1) is that there be some utterance-independent truth vehicle, a vehicle that possesses the value of truth during the era devoid of utterances. The natural choice of this vehicle would seem to be a proposition (conceived in accordance with some realist, rather than conceptualist or nominalist, theory of propositions); but one need not be committed to realism about propositions in order to ascribe truth to (1). One could, for example, postulate utterance-independent sentence-types, understood as universals that need not be instantiated by sentence-tokens in order to exist. According to this conception, (1) is true only if there is a formerly true sentence-type "The era devoid of linguistic utterances is present." Alternatively, one could (with Plantinga) postulate utterance-independent "states of affairs," so that (1) is taken as ascribing a past truth (or, if preferred, a past holding or obtaining) to a certain "state of affairs." Or again, one could postulate Platonistic classes or sets and interpret (1) as ascribing the relevant sort of membership to the relevant class.

Naturally, there are philosophical difficulties involved in each of these (and other) interpretations of (1). But it is not necessary to resolve, or even to mention, them in order to establish that (1) is true only if there is an utterance-independent truth vehicle expressed by the clause following "It was true that." If there are no utterance-independent truth vehicles and extreme nominalism is true, then (1) is false.

The necessary and sufficient truth conditions of (1), stated in a tenseless language, are that it is true if and only if three conditions hold:

a. There (is) an era devoid of utterances earlier than the utterance of (1).
b. There (is) some utterance-independent truth vehicle V that is expressed by the clause prefixed by "It was true that."
c. V (is) true during the era devoid of utterances.

Notice that my account of the truth conditions of (1) presupposes neither that events possess A-properties nor that (1) or the truth vehicle V refers to events with A-properties. Although I believe that these tenseless truth conditions, along with other true premises, entail that there are events with A-properties and truth vehicles that designate them (see chapter 4), I am not inferring truth conditions a–c from these entailments. Rather, my account is based on observations of how sentences like (1) are used; from these observations, I gather that the above truth condition rules of usage are being obeyed.

It seems obvious that none of the four token-reflexive constructions (of Reichenbach, Smart, Mellor, and Searle) discussed in section 3.1 have the same truth conditions as (1); for all of these constructions translate "is present" by a phrase that predicates *simultaneity with an utterance* of the era devoid of utterances, resulting in a sentence that is true in no possible conditions. For example, Smart's rules of translation imply that (1) is translated by

(1A) It (is) true, earlier than this utterance, that the era devoid of utterances (is) simultaneous with this utterance

which is self-contradictory regardless of whether the second "this utterance" is interpreted as referring to an utterance of (1A) or some other expression. The situation is not improved if we follow Mellor or Searle and substitute for the phrase "(is) simultaneous with this utterance," "(is) simultaneous with the utterance U" or "(is) at *cotemporal," where "U" denotes an utterance of (1) and "at *cotemporal" expresses a Fregean sense that refers to the utterance that expresses it.

The token-reflexive theorists would undoubtedly not remain silent in face of this criticism. One objection some of them might make is that (1) can no more be uttered with truth than its corresponding token-reflexive sentence. It might be argued that utterances of "is present" always denote the time of their utterance, which in tensed language may be called "the speaker's present." To utter (1) is to assert that it was true that the utterance-free era occupies the speaker's present, the time at which (1) is uttered. But it is impossible that this could ever be true. Consequently, (1), like its token-reflexive translation, cannot be uttered with truth under any conditions and in this respect has similar truth conditions to the token-reflexive sentence.

Moreover, if (1) is reformulated as a sentence that can be uttered with truth, the objection continues, it will have the same truth conditions as its new token-reflexive translation. Reformulated, (1) becomes

(2) It was true that the era devoid of linguistic utterances was present.

Item (2) is true if and only if there is an utterance-free era earlier than the utterance of (2) and there (is) earlier than the utterance of (2) a truth vehicle with the value of truth that states or implies that the era devoid of utterances (is) earlier than the utterance of (2). But these are just the truth conditions of

(3) It (is) true, earlier than this utterance, that the era devoid of linguistic utterances (is) earlier than this utterance.

Thus, the objection concludes, A-sentences about truths about past or future times devoid of utterances have the same truth conditions as the relevant token-reflexive sentences about these truths.

I have two independent responses to this objection. The first is that the objection is based on a confusion of the A-word "now" with the A-phrase "is present." It is true that utterances of the A-word "now" always and only denote the speaker's present, even if the clause of which this word is a part is prefixed by a past or future tensed operator. But this is not true for utterances of "is present," which denote the speaker's present only if the clause of which this phrase is a part is not prefixed by a past or future tensed operator. Utterances of both

(4) The storm is occurring now

and

(5) It will be true that the storm is occurring now

refer to a storm that is occurring at the speaker's present. But utterances of

(6) The storm is present

and

(7) It will be true that the storm is present

have different temporal denotations. Each utterance of (6) entails that the storm is present at the speaker's present, but each utterance of (7) entails that the storm is later than the speaker's present. Thus, it might be said, "It will be true tomorrow that the storm is present—so start getting prepared"; but no one would say this if "occurring now" were substituted for "present." This property of the A-word "now" has been discussed in a wider framework by Hans Kamp and Castañeda,[16] and the difference between "now" and "is present" has been noted by Prior. The latter writes:

> In "It will be the case tomorrow that my sitting down is present," the presentness referred is a presentness that will obtain tomorrow, i.e. at the time to which we are taken by the tensing prefix. But in "It will be the case tomorrow that I am sitting down now," the word "now" indicates the same time that it would indicate if it occurred in the principal clause—the time of utterance.[18]

We may say correspondingly that the past-tense prefix of (1) takes us to the earlier time when the truth vehicle expressed by "the era devoid of linguistic utterances is present" was true, that is, to the time that *was* denoted by this truth vehicle when it *was* true. It does not take us to the time that *is now* denoted by this truth vehicle. What is now expressed by "The era devoid of linguistic utterances is present" now denotes the speaker's present; thus, this truth vehicle is now false. But that is perfectly consistent with saying that it was true. Indeed, that is all that is said by (1). (1) is consistent with both

(6) It is now false that the era devoid of linguistic utterances is present

and

(7) It is now false but was true that the era devoid of linguistic utterances is present.

The phrase "is present" in (1) takes us to the present that was denoted by the truth vehicle (rather than to the present that is now denoted by this truth vehicle), since it is prefixed by the operator "It was true that." In fact, that is precisely the function of this operator—to ascribe a past, rather than present, denotation to the truth vehicle expressed by the clause it operates upon.

These considerations show that (1) must be distinguished from

(8) It was true that the era devoid of linguistic utterances is occurring now

and that (1) is immune to the charge that can be leveled against (8), namely, that it is logically impossible to utter it with truth. They also show that the consistent token-reflexive sentence

(3) It (is) true, earlier than this utterance, that the era devoid of linguistic utterances (is) earlier than this utterance

will not do as a translation of (1). (3) is the token-reflexive sentence that corresponds to

(2) It was true that the era devoid of linguistic utterances was present

and that (according to the tenets of the token-reflexive theory) translates (2). But if (3) translates (2), it cannot translate (1), for (2) is a different sentence from (1). (2) ascribes a past truth to the truth vehicle expressed by "the era devoid of linguistic utterances was present," and (1) ascribes a past truth to the truth vehicle expressed by "the era devoid of linguistic utterances is present"; since these two truth vehicles are different (e.g., they cannot be simultaneously true),[19] the sentences that express them are different.

There is a second reason why (3) cannot be the token-reflexive translation of (1). It is a tenet of the token-reflexive theory that present tensed A-clauses such as "the era devoid of linguistic utterances *is present*" are translated by, and express the same truth vehicles as, token-reflexive clauses that ascribe the relation of *simultaneity* to the event or era referred to and the expressing utterance. This rules out (3) as a possible translation of (1), for (3) does not ascribe the relation of *simultaneity* but of *being earlier*.

This constitutes my first response to the objection that A-sentences about truths concerning utterance-free times have the same truth conditions as their corresponding token-reflexive sentences. My second response avoids the issue of whether or not utterances of "is present" always denote the speaker's present and proceeds by placing the clause "the era devoid of linguistic utterances is present" in quotation marks. This response may be phrased in terms of propositions. Consider

(9) Proposition p_1 was true during the utterance-free era. Proposition p_1 was not expressed during this era; but had it been expressed then, it would have been expressed by a token of "The era devoid of utterances is present."

With this compare the purported token-reflexive translation of both the quoted clause and the preceeding phrases:

(10) Proposition p_1 (is) true, earlier than this utterance, during the utterance-free era. Proposition p_1 (is) not expressed during this era; but had it been expressed then, it would have been expressed by a token of "The era devoid of utterances (is) simultaneous with this utterance."

The proposition expressible by "The era devoid of utterances (is) simultaneous with this utterance" is true only if the era devoid of utterances bears the B-relation of simultaneity to some utterance. But in no possible world does the era devoid of utterances contain an utterance. Therefore, the proposition expressible by this clause is true under no possible conditions. By contrast, the proposition expressible by "The era devoid of utterances is present" is true if and only if (or when and only when) it is simultaneous with the era devoid of utterances. Since in some possible world this proposition is simultaneous with this era, this proposition is true under some possible conditions and therefore is different from the proposition expressible by "The era devoid of utterances (is) simultaneous with this utterance." This difference between the two propositions entails that sentence (9) is true under some possible conditions and that (10) is not.

These reflections can be deepened and expanded if we consider a related issue that George Schlesinger raised concerning past and future times at which there are no utterances. He asks, "Is it possible that E should have occurred in the past although it is not the case that there was ever a time at which E was occurring in the present?"[20] and replies:

One might make out a case that Reichenbach and Smart are committed to the view that such a thing is possible. It could be claimed that if E occurred in the distant past when there were no sentient beings and no utterances, then E occurred at some time *t* where *t* was in the past (i.e. *t* occurred before this utterance), yet E had never been in the present, because it had never been the case that E was simultaneous with any utterance.[21]

The context of Schlesinger's question-and-answer section suggests that Schlesinger subscribes to the belief expressed in the answer, the belief that it is coherent to suppose that

(11) E is past, but E never has been present

and that (11) is translated by

(12) E (is) earlier than this utterance, but it has never been the case that E (is) simultaneous with any utterance.

Put in terms of truth conditions, this implies that

(13) E has been present

is not a necessary and sufficient condition of the truth of

(14) E is past.

I believe that (11), unlike (12), is self-contradictory and therefore cannot be translated by (12) and that (13) is a necessary and sufficient truth condition of (14). To
suppose otherwise is to hold something manifestly at odds with the truth condition
rules of usage of A-sentences like (14). By this I do not mean that this supposition is
at odds with my stipulative definition of these rules—or with some tenser's stipulative definition of these rules—but with the rules in fact followed by English-
speaking people. Schlesinger seems to hold that there is no matter of fact at stake
concerning these rules of usage and that there are only the varying definitions of
these rules stipulated by tensers or detensers. But this seems false. English-speakers
in fact use A-sentences like (14) in accordance with some truth condition rules of
usage, and these rules are open to empirical observation. That these observations
support my claim that (14) is true if and only if (13) is true is evinced in part by
sample conversations like this:

> s.p.$_1$: The volcanic eruption described by the speaker wasn't present at all.
>
> s.p.$_2$: You're mistaken. I clearly remember his saying that the volcano *had*
> erupted.
>
> s.p.$_1$: Of course, it *has* erupted; I grant it is past. All I'm saying is that it
> wasn't present.
>
> s.p.$_2$: What?!

The speaker s.p.$_2$'s startled response is warranted, since s.p.$_1$ violated the conventionally accepted truth conditions rules of usage of past tense A-sentences. But
consider how out of place a startled reaction would be in this conversation:

> s.p.$_1$: The volcanic eruption described by the lecturer has never been simul-
> taneous with any utterance.
>
> s.p.$_2$: I remember his saying that it was earlier than his lecture.
>
> s.p.$_1$: Of course, it was earlier than his lecture; all I'm saying is that the
> volcano had not been simultaneous with any utterance when it erupted.
>
> s.p.$_2$: I see.

This exemplifies the claim that there are matters of fact concerning the truth condition rules of usage of A-sentences, and that these facts lend support to the conclusion that A-sentences as they are normally used are untranslatable by token-reflexive
sentences.

3.3 The Confirmation Conditions of A-Sentences and Tenseless Token-Reflexive Sentences

In section 3.2 I considered A-sentences about truths about distant past or future time
at which there are no utterances. In the present section, I shall focus my attention on

A-sentences of a second sort that also prove especially troublesome to the token-reflexive theory of A-sentences, namely, A-sentences about the speaker's not uttering anything. I will study these sentences from the point of view of our second translation condition, that the two sentences be confirmed/disconfirmed by observations or experiences of the same sort.

Compare

(1) I am not uttering anything

with any one of its alleged token-reflexive translations, for example,

(2) I (am) not uttering anything simultaneously with this utterance.

If I am observed to utter (1) on some occasion, (1) will be disconfirmed. (Here I am using "utterance" in Smart's sense of a spoken token or the earliest time-slice of an inscribed token, this time-slice being the period during which the token is being written down.) But I need not utter (1), I could think it silently to myself—run the sentence through in my mind without speaking or inscribing it. As silently thought, it is true; and I can confirm it as true by experiencing or observing that I (am) not saying or inscribing anything while I (am) thinking it. But (2) is false whether I speak it, inscribe it, or think it silently to myself, since it is false a priori. Consequently, there are observations that would confirm (1) but not (2).

A different argument concerning sentences like (1) can be formulated if we assume a realist or conceptualist theory of propositions, according to which propositions can be entertained and/or exist at times when they are not being vocally or inscriptionally expressed. The argument is straightforward: the proposition that I might have expressed at some time t by some utterance of (1) is confirmed as true at t by observations that I am not uttering anything at t, but the proposition I might have expressed at some time t by some utterance of (2) is false a priori and has no possible observational confirmation.

This latter argument is somewhat related to an argument Castañeda uses to refute the token-reflexive theory of "I." Castañeda writes (using "statement" to mean proposition):

> Reichenbach, for instance, claims that the word "I" means the same as "the person who utters this token." This claim is, however, false. A statement formulated through a normal use of the sentence "I am uttering nothing" is contingent: if a person utters this sentence he falsifies the corresponding statement, but surely the statement might, even in such a case, have been true. On the other hand, the statements formulated by "The person uttering this token is uttering nothing" are self-contradictory: even if no one asserts them, they simply cannot be true.[22]

If we substitute "I (am) uttering nothing simultaneously with the utterance" for "The person uttering this token is uttering nothing," the same argument (beginning with "A statement formulated . . . ") applies to Smart's and others' token-reflexive account of A-expressions.

The radical difference between A-sentences about not saying anything and their corresponding token-reflexive sentences is further illustrated by their noninter-substitutivity in belief reports. The A-sentence

(3) John believes he is uttering nothing at the time he believes to be present

ascribes to John a possibly true belief; but its alleged token-reflexive translation,

(4) John believes he is uttering nothing at the time he believes to be simultaneous with his utterance,

ascribes to him a self-contradictory belief.

It is worthwhile noting at this point how my basic argument applies to Searle's explicitly propositional version of the token-reflexive theory of A-sentences. Searle would not translate "I am not uttering anything" by "I (am) not uttering anything simultaneously with this utterance" but by

(5) I (am) not uttering anything at *cotemporal.

"At *cotemporal" is used to express "Fregean senses" each of which refers to "the utterance in which it is expressed."[23] Each of these senses also refers to the time of its expression by the utterance, i.e., to whatever moment or set of events is simultaneous with the expressing utterance. Each sense also attributes the relation of simultaneity to this moment or set of events and the event that the rest of the sentence is about. Given this analysis of "at *cotemporal," it follows that the complete Fregean senses or propositions expressed by uses of (3) refer both to my silence and to my utterance of "at *cotemporal" and ascribe to them the relation of simultaneity. Such propositions cannot be confirmed by any possible observation, since they are self-contradictory and therefore differ from the proposition(s) expressed by (1).

George Schlesinger would remain unconvinced by all the arguments presented in this section. He first states the tenser's position:

> Another objection to the view that equates the phrase "at present" with the phrase "is simultaneous with this utterance" has been that this cannot be so because the proposition "At present nothing is being uttered" is a contingent proposition, whereas "At the time which is simultaneous with this utterance nothing is being uttered" is self-contradictory.[24]

Schlesinger then offers this response, to which he appears to subscribe: "Reichenbach and Smart do seem to be committed to the view that 'At present nothing is being uttered' is also self contradictory. If this strikes us as strange, it is only because we have been tacitly assuming a different analysis of 'at present'."[25] Schlesinger again seems to be implicitly expressing the view that there are no matters of fact at stake concerning the rules of usage of the relevant A-expressions that could enable one to decide between the two analyses. Schlesinger's adherence to this view explains his reliance in *Aspects of Time* and *Metaphysics* on arguments based on our attitudes and emotional experiences of time (e.g., our wish to be ten years younger or our dread of the future) to make a case for the tenser's position. I believe that Schlesinger's arguments concerning our experiences are both fruitful and plausible, but I do not share his apparent pessimism with regard to semantic arguments. I believe that semantic arguments on behalf of the tenser's position are at least as viable as the arguments based on our emotional experiences. An integral element in my belief is that there are established rules of usage of English that are open to empirical inspection and to which one can appeal to determine whether the tenser's or detenser's analysis is the correct one. For example, I believe that it is an established rule of English that what is stated by "At present nothing is being uttered" is logically contingent and thereby differs from what is stated by "At the time which is simultaneous with this utterance nothing is being uttered," which is necessarily false. My belief that this is a rule of English is based upon my observations of how such sentences as these are actually used. For example, it is an observable fact that many English users report (or would report if you asked them) that on some occasions they have silently thought to themselves such sentences as "At present nothing is being uttered" and believe such sentences to be confirmed on these occasions by their own and others' silence. Thus, if I "have been tacitly assuming a different analysis [from] Reichenbach's and Smart's," this is not because I am begging the question against them but because my analysis is based upon observations of how the relevant sentences are used by English speakers. My arguments rest on empirical observations no less than do the arguments based on our emotional experiences of time; the former arguments rest upon observations of linguistic behavior, the latter, upon observations of emotional behavior.

3.4 The Entailment Relations of A-Sentences and Tenseless Token-Reflexive Sentences

Our results in the last two sections suggest that A-sentences of a certain sort need not be uttered in order to be true and that the propositions (or other truth vehicles) expressed by them do not refer to utterances. In this section, these results will be generalized by applying them to all normal A-sentences. (By a "normal A-sentence" I mean one that is not a token-mentioning A-sentence such as "This utterance is present.") I shall argue that each normal A-sentence is logically nonequivalent to its alleged token-reflexive translation, since their entailment relations concerning their own utterances and expressed propositions are different.

In section 3.3 I indicated that "I am uttering nothing" is logically contingent, since it can be mentally tokened and be true, whereas "I (am) uttering nothing simultaneously with this utterance" is self-contradictory. This is the most glaring case of a difference that obtains between all normal A-sentences and their corresponding token-reflexive sentences; that is, normal A-sentences need not be uttered in order to be true, whereas the token-reflexive sentences must be uttered to be true. For example, it is possible silently to think to myself a true token of

(1) The forest is now burning;

but it is logically impossible silently to think a true mental token of its corresponding token-reflexive sentence,

(2) The forest (is) burning simultaneously with this utterance.

In terms of logical nonequivalence, this difference between (1) and (2) may be put by saying that (2) but not (1) entails

(3) There (is) some utterance about the burning of the forest that (is) simultaneous with this burning.

This difference between A-sentences and their corresponding token-reflexive sentences requires that the rules of usage of A-sentence-tokens discussed in chapter 2 be appropriately extended. In this chapter, I was confining myself more or less implicitly to physical tokens of A-sentences—their vocal productions and inscriptions. In terms of these physical tokens, it is truly said (for example) that each present tensed physical A-sentence-token entails that there is some utterance (in Smart's sense) that is simultaneous with the event that the sentence-token is about. But the recognition that A-sentences can be mentally tokened and be true requires that we extend these rules to nonphysical tokens, as well. Thus, we should say that the *A-sentence* (1) entails

(4) There (is) some sentence-token about the forest's burning that (is) simultaneous with this burning,

where "sentence-token" may refer to a physical or mental token. Since (4) is nonequivalent to (3) and since (1) entails (4) but not (3), (1) is nonequivalent to its alleged translation, (2).

The idea that each A-sentence is translated by a token-reflexive sentence that refers to an *utterance* is part and parcel of the "token-reflexive theory" espoused by Smart, Reichenbach, Kneale, Searle, and some others.[26] Their accounts of the translating sentences disallows the possibility that these sentences refer in some instances to *unuttered tokens* of sentences, to mental tokens. However, there is a different, "psychological" theory of A-sentences that hypothesizes that these sentences refer to psychical phenomena; and this theory may be taken as consistent with the assumption that some of these psychic phenomena are mental tokens. Accordingly, this psychological theory allows for the possibility that there are tenseless sentence-tokens referring to mental tokens and that these tenseless sentence-tokens correspond to the mental A-sentence-tokens. These tenseless sentence-tokens would be equivalent to the mental A-sentence-tokens in a way that Smart's and others "token-reflexive sentence-tokens" are not. But despite this advantage of the psychological theory, this theory is beset by a host of difficulties that render it unviable. I shall discuss these difficulties in section 3.6, after I have finished with the token-reflexive theory.

At this juncture, I shall turn to a second decisive difference between A-sentences and their corresponding token-reflexive sentences, a difference that concerns the rules of these sentences that make explicit reference to the truth vehicles expressed by them. In sections 3.2–3 I argued that A-sentences about times when there are no utterances and about my not uttering anything express truth vehicles (propositions if you prefer) that do not refer to utterances and thereby differ from their corresponding token-reflexive sentences, which do express truth vehicles that refer to utterances. The difference between these sorts of A-sentences and their corresponding token-reflexive sentences is especially noteworthy, since the A-sentences are consistent, whereas their corresponding token-reflexive sentences are not. But this difference still shows up for other A-sentences, even though their corresponding token-reflexive sentences do not differ so greatly as to be self-contradictory. The difference is that normal A-sentences express utterance-independent truth vehicles (truth vehicles that do not need to be expressed by utterances in order to be true), whereas no token-reflexive sentence expresses such truth vehicles. If we include mental sentence-tokens, as well, we can say that normal A-sentences—but no token-reflexive sentences—*express sentence-token-independent truth vehicles.*

Before I present evidence for this claim, it is necessary to clarify a certain matter, namely, that the rules concerning these truth vehicles do not imply that the nominalist theory of propositions is false. I noted in section 3.2 that these truth vehicles may be conceived as propositions, or sentence-types, or states of affairs, or sets, or something else. I prefer to conceive of them as propositions, but a nominalist may conceive of them otherwise. In my arguments, I usually refer to propositions, rather than some other sort of truth vehicle; but my arguments go through even if another sort of truth vehicle is postulated, instead (as long as the truth vehicle meets the conditions laid down by the semantic rules to which I am appealing).

The rules I have in mind include the rule that if a normal A-sentence is used on some occasion to express something true, what the A-sentence expressed on that occasion would have been true then even if it had not been expressed. For example, if at a certain time I say "The forest is now burning," what I express at this time would have been true then even if it were not then expressed by me (or anybody else). Evidence that this is a rule of normal A-sentences is that it is tacitly taken for granted in ordinary and scientific conversations that normal A-truths are not dependent upon their linguistic expression, and assertions that they are so dependent result in violations of linguistic conventions. Consider for example this conversation:

s.p.$_1$: Is it true that the forest is now burning?

s.p.$_2$: Yes, but it wouldn't have been true if you hadn't just asked that question.

s.p.$_1$: What?!

The startled response of s.p.$_1$ is due to the fact that s.p.$_2$ has violated the conventional presumption that the truth or falsity of what is said by "the forest is now burning" is not dependent (logically or empirically) upon whether or not this sentence is tokened. But there is no startled response in this conversation:

s.p.$_1$: Is it true that the forest is burning simultaneously with my utterance of this sentence?

S.P.₂: Yes, but it wouldn't have been true if you hadn't just asked that question.

S.P.₁: I see.

The implicit presumption that what is expressed by normal A-sentence is token-independent is also evinced by such conversations as these:

S.P.₁: It has been true for some time now that Melee Gatorp is acting strangely.

S.P.₂: I don't believe it. There is no evidence that anybody before you has spoken, written, or thought to themselves the sentence "Melee Gatorp is acting strangely."

S.P.₁: What?!

S.P.₁: If my calculations are right, then it will soon be true that the velocity of recession of the universe is less than 10^{-3}.

S.P.₂: Only if somebody will be around to say, "The velocity of recession of the universe is less than 10^{-3}."

S.P.₁: What?!

S.P.₁: It is true that Saul Bellow is still alive.

S.P.₂: If you hadn't said that, it wouldn't have been true that Saul Bellow is still alive.

S.P.₁: What?!

Notice that S.P.₂'s responses would not count as violations of linguistic conventions if the sentences uttered by S.P.₁ were token-reflexive. "That wouldn't have been true if you hadn't said that" is a perfectly appropriate response to "It is true that Bellow is alive simultaneously with this utterance." This suggests that there is a significant difference between the rules governing the truth vehicles expressed by normal A-sentences and the rules governing the truth vehicles expressed by token-reflexive sentences. In terms of logical equivalence, it may be said that the rules imply that what is expressed by normal A-sentences does not have the same truth value in all possible circumstances as what is expressed by their corresponding token-reflexive sentences. For example, what is expressed at t by "The forest is now burning" does not have the same truth value in all possible circumstances as what is expressed at t by "The forest (is) burning simultaneously with this token." Let us name "p_1" what is expressed at t by the A-sentence and "p_2" what is expressed at t by the token-reflexive sentence. If both p_1 and p_2 are in fact true when expressed at t, it is nevertheless the case that p_1 would have been true at t even if it had not then been expressed by any physical or mental sentence-token, whereas p_2 would not have been true at t if it had not then been expressed by any sentence-token. In possible worlds terminology, we may say that in all possible worlds similar to our own except in that p_1 and p_2 are not expressed at t, p_1 is true at t in those worlds, but p_2 is not.

This difference between A-sentences and token-reflexive sentences is perfectly compatible with several similarities between them. Each present tensed A-sentence-token is true if and only if it is simultaneous with the event it is about, is confirmed by observations of this simultaneity, and entails that there is some sentence-token simultaneous with the event it is about. Exactly the same holds true for the tokens of the token-reflexive sentence that corresponds to the A-sentence. But whereas the similarities between these two sentences lie in the semantic rules about the *tokens* of these sentences, the differences between them lie in the rules about *what is expressed* by these tokens. What is expressed by a token of the token-reflexive sentence is true if and only if it is expressed by a token that is simultaneous with the event that the truth vehicle is about. The truth vehicle expressed is confirmed by observations of the simultaneity between this token and the event—and by no other observations. And the truth vehicle expressed entails that there is some token simultaneous with the event that the truth vehicle is about. But none of these semantic characteristics apply to the truth vehicles expressed by tokens of normal A-sentences. What is expressed by a token of a present tensed normal A-sentence is true *if* it is expressed by a token that is simultaneous with the event that it is about but not *only if* it is expressed by such a token. It may be true even if it is not expressed by any token. (But *if* it is expressed by a token, it is true *only if* the expressing token is simultaneous with the event.) Moreover, what is expressed is confirmed by observations of the simultaneity between its expressing token and the event; but such observations are not required. If no tokens are observed, the truth vehicle may be confirmed by evidence that it itself is simultaneous with the event that it is about (see chapter 4). Finally, what is expressed by the A-sentence-token does not entail that there be some token simultaneous with the event; even if the truth vehicle is in fact expressed by such a token, there are some possible worlds in which the truth vehicle is true even though there are no tokens simultaneous with the event.

These semantic differences between what is expressed by tokens of these two sentences is reflected in corresponding differences between the sentence-tokens themselves. These differences appear when these tokens are not described in abstraction from what they express but *as expressing truth vehicles*. For example, the A-sentence-tokens and the token-reflexive sentence-tokens have different entailment relations concerning what they express. To return to our earlier example, we may say that each token of "The forest is now burning" entails

(5) There (is) something sententially expressed about the burning forest whose truth value (is) logically and empirically independent of its being sententially expressed.

But no token of the corresponding token-reflexive sentence "The forest (is) burning simultaneously with this token" entails (5). This may be taken as a manifestation of the logical nonequivalence of these two sentences.

So far, in this section, I have been concerned with showing the logical nonequivalence of normal A-sentences and their corresponding token-reflexive sentences. But this concern with normal A-sentences is not meant to suggest that I believe all other token-mentioning A-sentences, such as "This utterance is present," are logically equivalent to their corresponding tenseless token-reflexive sentences. On the contrary, I believe that sentences of these two sorts are manifestly non-equivalent. Compare, for example,

(6) This sentence-utterance is present, but it might not have been

with its corresponding tenseless token-reflexive sentence,

(7) This sentence-utterance (is) simultaneous with this utterance, but it might not have been.

Each utterance of (6) is true, since (6) might not have been uttered at any time that it is, in fact, uttered. But each utterance of (7) is false, since it is logically impossible for any sentence-utterance *not* to be simultaneous with itself. Such examples concerning token-mentioning A-sentences are sufficient by themselves to expose the inadequacy of the token-reflexive theory of A-sentences.

3.5 The Logical Identities of A-Sentences and Tenseless Token-Reflexive Sentences

Paul Fitzgerald presents an argument in "Nowness and the Understanding of Time" that, if sound, enables a valid demonstration of the logical nonidentity of A- and token-reflexive expressions to be constructed upon the very premises of Smart's token-reflexive translation theory. He observes:

> The analysandum
>> Event E is occurring now
>> contains five words. If it is implicitly self-referring, as the suggested analysis implies, then it refers to a five-worded token. But the analysans, namely
>> Event E occurs simultaneously with this token
>> refers to a seven-worded token. Since the token to which the analysans refers is different from the token to which the analysandum refers, and since neither of these tokens is a logical construction, the analysans and the analysandum are not logically equivalent. The idea here is that the truth of one token demands the existence of an entity (that token itself), whose existence is not demanded by the truth of the other token.[27]

This argument can be used to show that the tensed token and its alleged token-reflexive translation do not refer in the same way to the same item or do not ascribe to this item the same property or properties. If the tensed token, call it T_1, refers to itself, then it refers to an item different from that of the tenseless token-reflexive token; for the latter refers not to T_1 but to T_2—T_2 being the tenseless token-reflexive token. Therefore, the two tokens are logically different and cannot translate each other.

This implies that Smart's theory that A-tokens are translated by tenseless token-reflexive tokens is self-contradictory; for if it is true, then two contradictory assertions must be true, namely:

(1) A-tokens refer to themselves, that is, A-tokens.

(2) A-tokens are translated by, and thus have the same reference as, token-reflexive B-tokens; that is, they refer not to A-tokens but to B-tokens.

Schlesinger believes that Smart's token-reflexive theory can be rescued from this dilemma: "Fitzgerald misses the point of the analysis, which is to instruct us that the term 'now' is to be treated as an abbreviation and is itself to be taken to stand for 'simultaneously with this token'. Thus 'this token' refers to the original five-worded token."[28] Since both the A-token and the B-token refer to the A-token, the A-token and its translation refer to the same item, and the contradiction is resolved.

But Schlesinger's claim cannot be correct. If "now" is an abbreviation of "simultaneously with this token," such that tokens of the latter expression refer not to the sentence-tokens of which they are parts but to the sentence-tokens of which tokens of "now" are parts, then "now" and "simultaneously with this token" must obey the principle of the extensional interchangeability of complete expressions and their abbreviations. This principle is a necessary but insufficient condition for any expression a to be an abbreviation for some expression b:

P: For any abbreviation a and for any expression b of which a is an abbreviation, if a token of a is uttered in some extensional context, then a token of b might have been uttered in place of the token of a such that *either* the token of b would possess the same truth value in fact possessed by the token of a *or* (if the tokens of a and b are not sentence-tokens but parts of sentence-tokens) the sentence-token of which the token of b is a part would possess the same truth value in fact possessed by the sentence-token of which the token of a is a part.

For example, suppose in a certain circumstance I utter a token of "Einstein's STR was first published in 1905"; it follows from P that if I had, instead, uttered in that circumstance a token of "Einstein's Special Theory of Relativity was first published in 1905," the latter utterance would have had the same truth value as the former utterance.

But now suppose that in a certain circumstance X, I utter a true token T_1 of "The hurricane is occurring now." If Schlesinger's theory is correct, then by P I might, instead, have uttered in X a token T_2 of "The hurricane (is) occurring simultaneously with this token," and T_2 would have had the same truth value as T_1. But if T_2 does not refer to itself but to T_1 (as Schlesinger maintains), it is false, since if I had uttered T_2 instead of T_1 in X, then T_1 would not have occurred and thus would not have been simultaneous with the hurricane. Therefore, "now" cannot be an abbreviation for "simultaneously with this token" where tokens of the latter expression are not understood to refer to the sentence-tokens of which they are parts but to the sentence-tokens of which tokens of "now" are parts.

The conclusion is that either the alleged token-reflexive translation refers to itself and thus is not a translation of the A-token (since it has a different reference from the A-token) or it refers to, and is abbreviated by, the A-token, in which case it also is not a translation of the A-token (since it fails to obey the principle of the extensional intersubstitutivity of abbreviations and the abbreviated). Smart's theory that A-expressions have token-reflexive translations cannot be saved.

Similar considerations show that Reichenbach's and Searle's theories fail to meet the translation condition of logical identity. Reichenbach's alleged translations involve expressions such as "the time at which this token is uttered," whereby similar problems arise about the reference of "this token." Searle's alleged translations fail because an utterance U of "The hurricane (is) occurring at *cotemporal" refers to U and thereby has a different reference from the utterance T_1 of "The hurricane is occurring now." (Mellor's theory will be discussed shortly.)

A defender of Smart's, Reichenbach's, and Searle's theories might respond by abandoning the translation claim but maintain that A-sentence-tokens are nonetheless token-reflexive and predicate only B-relations. It is possible for a detenser to hold consistently both that A-sentence-tokens refer to themselves and predicate B-relations between themselves and the events that they are about and that these sentence-tokens are untranslatable by *other* token-reflexive sentence-tokens, namely, the tenseless ones. The purpose of introducing the tenseless token-reflexive sentences that correspond to the A-sentences, the detenser might continue, is not to provide translating sentences that are logically identical to the A-sentences but to provide sentences that are *logically analogous* to the A-sentences. Tokens of

(3) The hurricane is occurring now

are logically analogous to tokens of

(4) The hurricane (is) occurring simultaneously with this token

in that they refer to analogous items—to *themselves,* to *tokens.* Apart from this logical analogy, they are logically identical; for tokens of each refer to the same event, the occurrence of the hurricane, and predicate the same dyadic property, the relation of simultaneity. The introduction of the corresponding sentence (4) allows one to see clearly the logical structure of (3), for (4) overtly displays its token-reflexivity and thus enables one to understand (by analogy) the token-reflexivity that is covertly displayed in (3). By formulating the correspondence relation between (3) and (4) in terms of logical analogy, rather than logical identity, the theory that A-sentences are token-reflexive can be preserved, even if the original translatability thesis must be dropped.

Although this reformulation of the token-reflexive theory has not, in fact, been adopted or proposed by the token-reflexive theorists, it is preferable to the original version of the theory inasmuch as it overcomes the difficulties discussed in this section. Nevertheless, the reformulation of the token-reflexive theory succumbs to a serious objection that also applies to the original version of the theory. Although I have not previously introduced this objection, I believe it is a powerful one and serves as a fitting way to conclude my criticism of the token-reflexive theory. The objection is that both the original and reformulated version of the token-reflexive theory *are inconsistent with the semantic facts about synonymous A-sentences.* Synonymous A-sentences are logically identical and logically equivalent, but this fact is inexplicable if these sentences merely ascribe B-relations between their tokens and the events they are about. The sentence

(3) The hurricane is occurring now

is logically equivalent to

(5) The hurricane is now occurring.

This entails that *what is expressed* by (3) is logically equivalent to *what is expressed* by (5). In specific terms, this means that what is expressed at t in circumstance C by a token T_1 of (3) is logically equivalent to what is expressed at t in the same circumstance C by a token T_2 of (5). But suppose that these tokens express tenseless token-reflexive truth vehicles—propositions, let us say. The token T_1 of (3) expresses the proposition

(6) that the hurricane (is) occurring simultaneously with T_1,

and the token T_2 of (5) expresses the proposition

(7) that the hurricane (is) occurring simultaneously with T_2.

But clearly, (6) and (7) do not entail each other. Surely, the hurricane could be simultaneous with T_1 without being simultaneous with T_2. The occurrence of a token of (3) does not *require* that (5) also be tokened, and the tokening of (3) simultaneously with the hurricane does not require that (5) also be tokened then. In the actual world, both T_1 and T_2 occur simultaneously with the hurricane; but there is some possible world W that is alike the actual world except that in W, T_1 occurs simultaneously with the hurricane, and T_2 does not. In W, the proposition (6) is true but the proposition (7) is not. Consequently, the suggestion that A-sentence-tokens express tenseless token-reflexive propositions must be rejected as incompatible with the semantic facts. A-sentences are not logically analogous to tenseless token-reflexive sentences in the manner claimed by the reformulation of the token-reflexive theory. They do not express propositions that *refer to their expressing tokens* and *attribute B-relations between the tokens and events.* They must express propositions of another sort.

I shall conclude this section by briefly indicating that these considerations also show Mellor's theory (in its revised or unrevised version) to be inadequate to the semantic properties of A-sentences. According to Mellor's theory, the token T_1 of (3) states only the fact (6) and the token T_2 of (5) states only the fact (7). But this cannot be the case; for in actuality, the fact stated by T_1 is logically equivalent to the fact stated by T_2, whereas the facts (6) and (7) are not logically equivalent. This can be shown in terms of one of the examples Mellor presents in *Real Time*. He writes, "[Let] S be any token of 'It is now 1980'. . . . S is true if and only if it occurs in 1980."[29] This token-reflexive fact is the only fact stated by S: "We have seen that using tensed sentences demands nothing more than knowledge of when they are true and when false, i.e., of their tenseless token-reflexive truth conditions; so that, I suggest, is all there is to the tensed beliefs expressed. The idea that tensed sentences also express non-token-reflexive beliefs in tensed facts . . . is a gratuitous and idle supposition."[30] But consider that a token S_1 of "It is now 1980" is uttered by John simultaneously with a token S_2 of "It is now 1980" that is uttered by Alice. It is an unimpeachable linguistic datum that the fact stated by S_1 is logically equivalent to the fact stated by S_2. But Mellor's theory fails to explain or be consistent with their logical equivalence, for his theory implies that the fact stated by S_1 is

(8) S_1 occurs in 1980

and that the fact stated by S_2 is

(9) S_2 occurs in 1980.

These facts are not logically equivalent; for Alice need not utter S_2 in order for John to utter S_1, and John need not utter S_1 in order for Alice to utter S_2. Accordingly, we must suppose that S_1 and S_2 express other facts, facts that are logically equivalent, if our theory is to be consistent with the linguistic data. These further facts, as I shall argue in chapter 4, are the facts involving the presentness of 1980. Both S_1 and S_2 state the fact *that 1980 has the A-property of presentness*.

3.6 The Psychological Theory of A-Sentences

Whereas the token-reflexive theory holds that A-sentences ascribe B-relations to events and *physical tokens,* the psychological theory holds that these sentences ascribe such relations to events and *mental phenomena* (sense data, awarenesses, images, etc.) that are experienced concurrently with the tokening of the A-sentence. Russell, for example, once held that the nowness of events is to be understood in terms of their simultaneity with sense data;[31] and Grünbaum analyzes "is present" in terms of simultaneity relations between events and experiences. Grünbaum writes:

> What qualifies a physical event at a time *t* as belonging to the present or as now is not some physical attribute of the event or some relation it sustains to other purely physical events; instead what so qualifies the event is that at least one human or other *mind-possessing* organism *M* experiences the event at the time *t* such that at *t*, *M is conceptually aware* of the following complex fact: that his having the experience of the event *coincides temporally* with an awareness of the fact that he has it at all.[32]

Other exponents of the psychological theory of A-expressions include Lynn Rudder Baker, C. J. Ducasse, R. B. Braithwaite, W. Salmon, and R. M. Blake.

The psychological theory enables some of the problems besetting the token-reflexive theory to be solved. For example, if the translation of

(1) The hurricane is occurring now

is

(2) The hurricane (is) occurring simultaneously with this sense datum,

the problems of translation discussed in section 3.5 do not arise. Tokens of (1) do not refer to themselves, and tokens of (2) do not refer to themselves, so there is no difference of reference in this regard. Moreover, tokens of (2) do not refer to tokens of (1), so neither does the problem about the failure of the intersubstitutivity *salva veritate* of tokens of (2) with tokens of (1) arise.

The psychological theory is also immune to one of the criticisms presented in section 3.4, namely, that A-sentences can be mentally tokened with truth but that token-reflexive sentences (as conceived by Smart and others) cannot. I can mentally token (2) with truth no less than I can mentally token (1) with truth. This criticism can also be deflected if the token-reflexive theory is synthesized with the psychological theory in the manner suggested in section 3.4, so that the translations of A-sentences are regarded as referring to "tokens" in the wide sense that includes both physical utterances and mental tokens. This synthesis is present in Mellor's theory. However, this synthetic theory is less preferable than the "pure" psychological theory, considering that the synthetic theory alone falls prey to all the criticisms of the token-reflexive theory presented in section 3.5.

Despite these two advantages of the pure psychological theory over the token-reflexive theory of Smart, Searle, and others, the psychological theory is not able to withstand relevantly modified versions of the other criticisms I have directed against the token-reflexive theory. If some of the sample sentences discussed in sections 3.2–5 are modified so that they are about experiences or mental phenomena, rather than physical utterances, the relevant arguments can be made to apply to the psychological theory. The principle sample sentence in section 3.2 can be changed to

(3) It was true that the era devoid of mental phenomena is present,

and the operated-on clause can be translated as "The era devoid of mental phenomena (is) simultaneous with this mental phenomenon." It is manifest that (3) is true under some conditions but that its psychological translation is true under no conditions; thus, an argument similar to the one presented in section 3.2 applies to the psychological theory.

The main sample sentence in section 3.3 can be modified to

(4) I am not experiencing anything,

which expresses a proposition that is confirmable by observations that I am not experiencing anything. However, "I (am) not experiencing anything simultaneously with this experience" expresses a proposition that cannot be confirmed by any possible observation.

Moreover, the rules of A-sentences that were discussed in section 3.4 and that show A-sentences to express truth vehicles that are token-independent also show these truth vehicles to be independent of psychological experiences. For instance, "That wouldn't be true if you weren't sensing anything" is an appropriate response to "Bellow is alive concurrently with this sensation"; but it is wholly inappropriate as a rejoinder to "Bellow is alive now."

Finally, it is to be observed that the proposition John expresses at t by "The hurricane is occurring now," while he is experiencing sense datum S_1, is logically equivalent to the proposition Sally expresses at t by "The hurricane is now occurring," while she is experiencing sense datum S_2; but the proposition

(5) that the hurricane (is) occurring simultaneously with S_1

is not logically equivalent to the proposition

(6) that the hurricane (is) occurring simultaneously with S_2.

These difficulties with the psychological theory of A-sentences suggest that a new semantics of A-sentences is needed, a *tensed* semantics. The development of this semantics is the task of the next chapter. It will then be shown that this tensed semantics can be extended to apply to all tenseless sentences, as well, and thus that presentism is true (see part II).

Notes

1. Reichenbach, *Elements of Symbolic Logic* (New York, 1947), 284.

2. William Kneale, "Is Existence a Predicate?," in *Readings in Philosophical Analysis*, ed. H. Feigl and W. Sellars (New York, 1949), 37.

3. J.J.C. Smart, "The River of Time," in *Essays in Conceptual Analysis*, ed. A.G.N. Flew (London: Routledge & Kegan Paul, 1966), 213–27; idem, *Philosophy and Scientific Realism* (London: Routlege & Kegan Paul, 1963).

4. Smart, *Philosophy and Scientific Realism*, 133–34.

5. D. H. Mellor, *Real Time* (Cambridge: Cambridge University Press, 1981), 42.

6. Ibid., 74.

7. Private communication of December 11, 1988. The earlier criticism is stated in my "Problems with the New Tenseless Theory of Time," *Philosophical Studies* 52(1987): 371–92.

8. Mellor, *Real Time*, 30–31, 42; my italics.

9. L. Nathan Oaklander, "A Defense of the New Tenseless Theory of Time," 1988 mimeograph.

10. Ibid., 7.

11. Mellor, *Real Time*, 77–78. I have italicized the occurrences of "tokens."

12. Oaklander, "Defense of the New Tenseless Theory," 8.

13. Mellor, *Real Time*, 76.

14. John Searle, *Intentionality* (Cambridge: Cambridge University Press, 1983), 228.

15. Ibid., 223.

16. Stephen Hawking and George Ellis, *The Large-Scale Structure of the Universe* 356.

17. Hans Kamp, "Formal Properties of 'Now'," *Theoria* 37(1971): 227–73; Hector-Neri Castañeda, "Indicators and Quasi-Indicators," *American Philosophical Quarterly* 4(1967): 85–100; idem, "The Semiotic Profile of Indexical (Experiential) Reference," *Synthese* 49(1981): 275–316.

18. A. N. Prior, "Now," *Noûs* 2(1968): 104.

19. In other words, they cannot be simultaneously true if "was present" is taken as elliptical for "was and is no longer present," rather than as elliptical for "was present and still is present." I am taking this phrase in the former way.

20. George Schlesinger, *Aspects of Time* (Indianapolis: Hackett, 1980), 111.

21. Ibid., 133.

22. Castañeda, "Indicators and Quasi-Indicators," 87.

23. Searle, *Intentionality,* 228.

24. Schlesinger, *Aspects of Time,* 110–11.

25. Ibid., 133.

26. Mellor allows that some tokens are mental, and so this criticism does not apply to him. See his *Real Time,* 37–38.

27. Fitzgerald, "Nowness and the Understanding of Time," *Boston Studies in the Philosophy of Science,* no. 20 (Dordrecht: Reidel, 1974), 267–68.

28. Schlesinger, *Aspects of Time,* 132.

29. Mellor, *Real Time,* 74.

30. Ibid., 87.

31. Bertrand Russell, "On the Experience of Time," *Monist* 25(1915): 212–33.

32. Adolf Grünbaum, *Modern Science and Zeno's Paradoxes* (Middletown, CT: Wesleyan University Press), 16–17.

4

The Tensed Theory of A-Sentences

4.1 The Transition to Positive Arguments for the Tensed Theory of Time

Chapters 1–3 began the argument for the tensed theory of time by developing an adequate method of linguistic analysis (the "translation method") and using it to demonstrate the falsity of the tenseless theory of A-sentences. Chapter 4 continues this argument by developing a tensed theory of A-sentences, that is, a theory stating that all A-sentences ascribe an A-property of presentness, present pastness, or present futurity.

The argument of chapters 2 and 3 was that the date-theory and token-reflexive theory of A-sentences were false, but this negative argument also produced some positive results concerning the tenseless rules of usage of A-sentences. In chapter 2, I explicated the *token-reflexive rules of usage* of A-sentences, rules implying that A-sentences are used to assert B-relations between the events denoted and the uses of the A-sentences. In chapter 3, I discussed the *token-independent rules of usage* of A-sentences, which imply that these sentences are used to express propositions that do not refer to their expressing tokens. My aim in this chapter is to use these conclusions as premises of an argument that A-sentences express A-propositions and that events possess A-properties. (By an A-proposition I mean one that ascribes presentness, present pastness, or present futurity.) I shall argue that although these rules of usage prevent A-sentences from expressing B-propositions, they allow A-sentences to express A-propositions. Indeed, the hypothesis that A-sentences express A-propositions not only is consistent with these tenseless rules of usage but also explains why A-sentences obey them. Since there is no other hypothesis consistent with these tenseless rules of usage, I shall argue, it follows that the A-hypothesis is the true one.

4.2 Outline of the Argument That A-Sentences Express A-Propositions

In this section I outline and partly defend the argument that A-sentences express A-propositions. This argument is fully developed in sections 4.3–4.5.

The results of chapters 2 and 3 may be summarized by the sentence

(1) A-sentences do not express the same propositions that are expressed by tenseless date-sentences, token-reflexive sentences, or psychological sentences.

But it does not follow from (1) that

(2) A-sentences express A-propositions,

for (1) is consistent with both

(3) A-sentences express the propositions that are expressed by B-sentences of another sort than date or token-reflexive or psychological B-sentences

and

(4) A-sentences express B-propositions of a sort unexpressible by B-sentences.

(Here "B-proposition" means a logically contingent proposition that refers to an event and ascribes to it a B-relation but not an A-property.)

Not much argument is needed to reject the first of these alternative theses, (3); it is clearly implausible, and no detenser has ever defended it. Why this is so becomes apparent once (3) is explained. If a token of an A-sentence like "The moon is behind a cloud" is translatable neither by tokens of a date-sentence, nor by tokens of a token-reflexive or psychological sentence but is nevertheless translatable by tokens of some B-sentence, then it must be translated by tokens of a sentence that ascribe a B-relation to the event denoted by the A-token and to some relational term other than the A-token itself or the mental experience of the utterer of the A-token. However, there are no plausible candidates for this other relational term. This A-token manifestly cannot express the proposition expressed by tokens of sentences like

(5) The moon (is) behind a cloud simultaneously with the flight of that bird.

(6) The moon (is) behind a cloud simultaneously with the gathering of the members of the department.

(7) The moon (is) behind a cloud one day later than the thunderstorm over Charleston.

The token of the A-sentence is not translatable by tokens of any of these sentences; it is obvious that the A-token is not true if and only if the moon's being behind a cloud (is) simultaneous with the flight of a bird or a departmental gathering, or (is) one day later than a thunderstorm. None of these events could have occurred, and the A-token still could be true; for even if there is no bird flight, and so on, the

A-token still could occur simultaneously with the moon's being behind a cloud. Moreover, the B-relation of the moon's being behind a cloud to these other events does not suffice to make the A-token true; for the B-relation could obtain even if the A-token is earlier or later than the cloud's obscuration of the moon.

Such reflections make it evident that if A-tokens are not translatable by tokens of date-sentences, token-reflexive sentences or psychological sentences, then they are not translatable by any sort of B-sentence.

The second possibility, (4), that A-sentences express B-propositions of a sort unexpressible by B-sentences, was implicitly shown to be unviable in section 3.5. I there argued that the propositions that A-sentences express cannot be B-propositions that refer to their expressing tokens, since if they did express such propositions, the logical equivalence of synonymous A-sentences would be inexplicable. Moreover, the idea that the propositions expressed by A-sentences refer to tokens is incompatible with the token-independent nature of these propositions that was made manifest in sections 3.2–5.

Apart from these B-propositions that refer to the A-tokens by which they are expressed, there are no other candidates for the category of "B-propositions that are expressible only by A-sentences." No philosopher has ever put forward such a candidate, and I believe that no such candidate is conceivable.

If (3) and (4) are false, as I have just argued them to be, then the way is open to add to

(1) A-sentences do not express the same propositions that are expressed by tenseless date-sentences, token-reflexive sentences, and psychological sentences

the suitable premises that enable

(2) A-sentences express A-propositions

to be deduced. The argument is

(8) A-sentences do not express the same propositions that are expressed by tenseless date-sentences, token-reflexive sentences, or psychological sentences.

(9) A-sentences do not express the propositions expressed by any other sort of B-sentence.

(10) A-sentences do not express B-propositions of a sort unexpressible by B-sentences.

(11) Therefore, A-sentences do not express B-propositions.

(12) Besides B-propositions, the only other candidate for the propositions expressed by A-sentences are A-propositions.

(13) Therefore, A-sentences express A-propositions.

I will shortly explain that the implication of (12), that A-propositions are suitable candidates for the propositions expressed by A-sentences, has not yet been demonstrated and requires some argumentative support. But first I shall show how "Events possess A-properties" can be deduced from sequence (8)–(13), given some additional premises. The crucial additional premise is that *some A-propositions are true*, which is deduced as follows:

(14) Some tokens of A-sentences are true. (E.g., there are true tokens of "I am thinking," "John will die," "The meeting is starting now," "March 1, 1904 is long past," and "The density of the universe used to be greater than its present density.")

(15) Sentence-tokens are true if and only if the propositions they express are true.

(16) Therefore (by 13), some A-propositions are true.

(17) Since A-propositions ascribe A-properties to events, A-propositions are true only if events possess A-properties.

(18) Therefore, events possess A-properties.

Note that the conclusion (18) cannot be watered down by a B-theorist to mean that events possess *mind-dependent* A-properties. The idea that A-sentences express propositions that refer to, or entail, the existence of minds or language users has already been rejected in chapter 3. A symbol of this fact is that the A-sentence "It used to be, but is no longer, true that the era devoid of minds is present" is so far from being self-contradictory that it is actually true. (See chapter 3 for detailed arguments.) The crucial issue—the one not fully addressed yet—is whether the non-mind-entailing propositions expressed by A-sentences *are A-propositions*. In this section and in the preceding chapters I have presented what may be styled a *negative* argument for (13), that A-sentences express A-propositions, an argument by *exclusion*. I have excluded B-propositions from the list of viable candidates for the propositions expressed by A-sentences but have offered no positive arguments to the effect that A-propositions are viable candidates for these propositions. I have shown that the tenseless rules of usage of A-sentences are incompatible with the hypothesis that these sentences express B-propositions, but I have not shown that these rules of usage are compatible with the hypothesis that they express A-propositions. The demonstration of this compatibility is necessary, since it does automatically follow that A-sentences express A-propositions if they do not express B-propositions. It is arguably conceivable that A-sentences express neither A-propositions nor B-propositions. This would, in fact, be the case if the tenseless rules of usage of A-sentences were inconsistent both with the hypothesis that A-sentences express B-propositions and with the hypothesis that they express A-propositions. Of course, beside A-propositions and B-propositions, there are no other candidates for the propositions expressed by A-sentences. At least, I am unable to conceive of any third candidate; and nobody else has ever put forward a third candidate.[1] If the tensed rules of use of A-sentences cannot be accounted for either by the A-theory or by the B-theory, we would we warranted in believing that

A-sentences do not express any propositions. Rather, they would express something else—perhaps fragments of propositions or nonsensical concatenations of ideas or images. The seemingly obvious "fact" that there are true tokens of A-sentences would turn out to be an illusion. A-sentence-tokens would lack truth value, and it would be nonsensical to report, "I *am* thinking," "The sun *was* setting over the mountains," or "The universe *will* expand to a greater radius." Complete skepticism—or, rather, complete *nihilism*—regarding the objective correlates of A-expressions would be warranted. Virtually all sentence-tokens in ordinary language and many of those in scientific language would have to be regarded as incoherent.

These considerations indicate that it is a necessary task to *demonstrate* that A-propositions are viable candidates for the propositions expressed by A-sentences. My aim in the following three sections is to show that the theory of A-propositions is both consistent with, and able to explain, the tenseless rules of usage of A-sentences. I first make it manifest that the A-proposition theory is both compatible with and provides an explanation for the tenseless token-reflexive rules of A-sentence-tokens. These are the rules stating the semantically essential B-relations that A-sentence-tokens sustain to the events they are about, which formed the topic of chapter 2. Following this, I show (in section 4.4) that the hypothesis that A-sentences express A-propositions is both compatible with, and explains, the tenseless token-independent rules governing the propositions expressed by A-sentence-tokens. These are the rules implying that the propositions expressed by A-sentence-tokens do not refer to their expressing tokens, which were the subject of chapter 3. Finally, in section 4.5, I develop a theory of A-propositions that is compatible with, and explains, the linguistic data concerning the similarities and differences between A-sentences that contain temporal indexicals such as "now" (e.g., "The meeting is starting now") and A-sentences that contain only tensed copulae or verbs (e.g., "The meeting is starting").

4.3 The Tensed Explanation of the Tenseless Token-Reflexive Rules of A-Sentence-Tokens

In chapter 2, I indicated that A-sentence-tokens are true if and only if they bear the appropriate B-relation to the event they are about. If A-sentence-tokens express A-propositions, then they have *tensed* truth conditions, as well; the tokens ascribe A-properties to events and are true if and only if the tokens and events possess the appropriate A-property. For example, any token T of "The storm is approaching" is true if and only if *T is present* and *the approach of the storm is present*. The first requirement in elucidating the tensed truth conditions is to show how they are consistent with the tenseless truth conditions; for it seems at first glance that if these tokens have tenseless truth conditions, it is impossible that they also have tensed truth conditions. If the tenseless truth conditions of these A-sentence-tokens are both necessary and sufficient to make them true, then how can any other conditions be either necessary or sufficient to make them true? A detenser argument would be that the occurrence of a token T_1 of "The storm is approaching" simultaneously with the approach of the storm is both necessary and sufficient to make T_1 true; therefore, *by definition* no other condition, such as the presentness of the storm's approach, is either necessary or sufficient to make T_1 true.

The response to this argument will provide us with an understanding of tensed truth conditions. First, we must distinguish between tenseless and present tensed predications of truth. To predicate truth tenselessly of some token T_1 is to say or imply that T_1 (is) true when it occurs, and to predicate truth in a present tensed manner is to say that T_1 is true where the "is" is present tensed. T_1's tenseless truth conditions are both necessary and sufficient for its tenseless truth but not for its present tensed truth. (Hereafter, "present tensed truth" will be expressed by "tensed truth," "present tensed truth conditions" by "tensed truth conditions.") Suppose that the token T_1 of "The storm is approaching" occurred two months ago, simultaneously with the storm's approach. Since the tenseless truth condition, *T_1's being simultaneous with the approach of the storm,* is met, T_1 (is) true when it occurs. But it is false that T_1 *is* (present tensed) true. Rather, T_1 *was* true, namely, when the storm was approaching. In order for the tenseless truth condition to render T_1 now true, another condition must be added, namely, the tensed condition that T_1 and the storm's approach are both *present.* The tenseless truth condition is necessary but insufficient for T_1's tensed truth. Unless T_1 (is) true when it occurs, it can never be the case that T_1 *is* occurring and *is* true; but its (being) true when it occurs is insufficient for T_1 to be now true, since "T_1 (is) true when it occurs" does not entail "T_1 is now occurring and is now true."

On the other hand, T_1's tensed truth condition, although sufficient for T_1's tenseless truth, is not necessary for it. If T_1 and the storm's approach are not present, T_1 may still be tenselessly true; for T_1 may be past but simultaneous with the storm's approach. T_1's tenseless truth does not require T_1's tensed truth (that T_1 be now true); but it does require that T_1 possess tensed truth *at some time,* that is, that T_1's possession of truth either was present, is present, or will be present. T_1's tensed truth conditions are sufficient for T_1's tenseless truth, since "T_1 and the storm's approach are both present" entails "T_1 (is) true when it occurs," as is made explicitly obvious below.

The conception of T_1's tensed truth and tensed truth conditions I have introduced is not the only possible one; and I shall mention two other conceptions at the end of this section and provide an argument for the soundness of the conception I am advancing. But first, let me develop it further.

My talk of "the present tensed truth and the tenseless truth of T_1" might be misleading, inasmuch as it may suggest that T_1 and other A-tokens have two truth values. In fact, there is only one truth value possessed by T_1, a value whose intransient temporal relations to other phenomena are describable in a tenseless language and whose transient temporal properties are describable in tensed terms. If T_1 and the storm's approach are both occurring now, then T_1 is now true, and the truth T_1 now possesses is not a different truth value from the one it possesses when it occurs. It is the same truth value considered on the one hand in respect of its A-position and on the other hand in respect of its B-position. "T_1 is true" in the present tensed sense describes T_1 as presently possessing the truth value, and "T_1 (is) true when it occurs" describes T_1 as possessing the truth value at the B-time of its occurrence. To say that T_1 can be true (tenselessly) without being true (in the present tensed sense) is to say not that T_1 can possess one truth value (a tenseless one) without possessing another (a present tensed one) but that T_1's truth can be truly predicated of it tenselessly at times when it cannot be truly predicated of it in a present tensed manner. If T_1 occurred with truth in February, then in April, it (is) true that *T_1 (is) true when it occurs* (in February), but in April, it is false that *T_1 is presently true*. In April, it is true that T_1 was true.

In the preceding pages I have argued that the possession of tensed truth conditions by A-sentence-tokens is consistent with their possession of tenseless truth conditions. This is the first step in providing a tensed account of the semantics of A-sentence-tokens. The second step is to make explicit that the tensed truth conditions *explain* the tenseless truth conditions. It can be made briefly manifest that the tenseless truth conditions can be deduced from the tensed ones but that the tensed ones cannot be deduced from the tenseless ones and that the tenseless truth conditions are in this sense "reducible" to the tensed ones. The tensed truth conditions of T_1 are the following:

(14) T_1 is present.

(15) The storm's approach is present.

The conjunction of (14) and (15) entails

(16) T_1 (is) simultaneous with the storm's approach,

which is the tenseless truth condition of T_1.

The tensed conditions of the present tensed truth of past and future tensed A-sentence-tokens also explain the tenseless truth conditions of these tokens. A token T_2 of "The storm was approaching" is true at present if and only if

(17) T_2 is present

and

(18) The storm's approach is past,

which jointly entail T_2's tenseless truth condition, that is,

(19) T_2 (is) later than the storm's approach.

Not only are the tenseless truth *conditions* of A-sentence-tokens "reducible" to their tensed truth conditions, but their tenseless *truth* is "reducible" to their tensed truth.

(20) T_1 is now true

entails, but is not entailed by,

(21) T_1 (is) true when it occurs.

Likewise, the past and future tensed truth of A-sentence-tokens entails, but is not entailed by, their tenseless truth; for

(22) T_1 was true

entails (21) no less than does (20) and, like (20), is not entailed by (21).

So far, I have discussed the truth conditions of A-sentence-tokens. I now want to add that the tenseless *confirmation conditions* of these tokens are also reducible to their tensed confirmation conditions. The token T_1 of "The storm is approaching" is confirmed to be presently true by observations that T_1 and the storm's approach are both present. The observation sentence

(23) There is observational evidence that T_1 and the storm's approach are both present

entails T_1's tenseless confirmation sentence,

(24) There (is) observational evidence that T_1 and the storm's approach (are) simultaneous;

but (24) does not entail (23).

The tenseless token-reflexive *entailment relations* of A-sentence-tokens discussed in chapter 2 concern B-relations between tokens and events. It was implied, for example, that T_1 entails

(25) There (is) some sentence-token about the storm's approach that (is) simultaneous with this approach.

Item (25) is deducible from the tensed token-reflexive entailment relations of T_1. T_1 entails

(26) T_1 and the storm's approach are both present,

which in turn entails, but is not entailed by, (25). That T_1 entails (26) is evident from the fact that T_1 cannot be presently true unless T_1 and the storm's approach are both present; in possible worlds terminology, at the time in each world at which T_1 is true in the present tensed sense, both T_1 and the storm's approach are present, which is to say that (26) is true at each time in each world in which T_1 is true.

In chapter 2 I indicated that a token-reflexive denotation is a part of the *logical identity* of A-sentence-tokens. A-sentence-tokens refer to themselves and predicate a B-relation between themselves and the event they are about. But in order to understand precisely in what sense this is true, a distinction must be made between *explicit* logical identity and *implicit* logical identity. The logical identity of a sentence is determined by what it refers to, how it refers to this item (directly or via a description), and what it predicates of this item. A sentence explicitly refers to an item if it contains a word or phrase that stands for that item, and it explicitly predicates a property if it contains a word or phrase that stands for that property. For example, "the sun has a shape" contains a phrase, "the sun," that stands for *the sun* and a phrase, "has a shape," that stands for the property of *having a shape*. Thus, this sentence explicitly refers to the sun and explicitly predicates of the sun the property of having a shape. This is the explicit logical identity of "the sun has a shape." This sentence also implicitly refers to, or predicates, something if it does not contain a word or phrase that stands for the item or property but nevertheless entails a sentence that does contain such words or phrases. "The sun has a shape" entails "the sun has a size," which explicitly predicates *having a size*. We may say, then, that "the sun has a shape" implicitly predicates of the sun the property of having a size and that this predication belongs to the implicit logical identity of "the sun has a shape." "The sun has a shape" and "the sun has a size" are nevertheless logically different from each other, considering that their explicit logical identities are different.

Given this distinction, it may be said that the reference of A-sentence-tokens to themselves and their predication of B-relations of themselves and events belong to the implicit logical identity of these sentence-tokens. The token T_1 of "the storm is approaching" does not contain a word or phrase—such as "this token" or "T_1"— that stands for itself and therefore does not explicitly refer to itself. Nor does it contain a word or phrase that stands for a B-relation; according to the A-theory, the present tense of the copula stands for the A-property of presentness. However, T_1 entails a sentence that explicitly refers to T_1 and explicitly predicates a B-relation, namely,

(16) T_1 (is) simultaneous with the storm's approach.

In general, we may say that the explicit logical identity of A-sentence-tokens is to refer to a thing or event and predicate an A-property; their implicit logical identity is to refer to themselves and predicate a B-relation. Their implicit logical identity, which may be called their "token-reflexive identity," is reducible to their explicit logical identity, their "A-identity," in the sense that A-tokens such as T_1 entail B-sentences such as (16) but are not entailed by B-sentences such as (16); that is, their token-reflexive identity is derivable from their A-identity, but their A-identity is not derivable from their token-reflexive identity.

This account of the tensed truth values and tensed truth and confirmation conditions, and so on of A-sentence-tokens may be deepened and defended by comparing it with two other accounts, one offered by D. H. Mellor and one defended by Graham Priest in their recent interchange of articles in *Analysis*.[2] Let me first restate my account of tensed truth conditions in terms of an example discussed by Mellor and Priest in order to provide a ready point of comparison:

(17) The token U of "John is dead" is true if and only if U and John's death (i.e., his state of being dead) are both present.

My account (but not Mellor's or Priest's) is token-reflexive, that is, includes an A-property of the token among the truth conditions of the token. Mellor advances a conception of tensed truth and then rejects it in favor of the idea that truth is only tenseless. According to Mellor's conception of tensed truth,

(18) The token U of "John is dead" is true if and only if John's death is present (and this is the only nontenseless truth value this token possesses).

Item (18), but not (17), allows that U is true at times when it is not present. If U is past or future and John's death is present, then U is (now) true. According to (17), if U is past or future, it cannot be now true, since U's presentness is a necessary condition of its present truth. One reason to favor (17) over (18) is that the latter has unpalatable consequences, as Mellor points out himself. (But Mellor is not aware of [17] as an alternative to [18] and believes that the only alternative is the conception of truth as solely tenseless.) Mellor points out that (18) implies that if U is uttered before John's death, then its tensed truth value is false but that this value changes to true after John dies. "But it is absurd to say . . . that John's death posthumously verifies every premature announcement of it."[3] Mellor concludes that coherency requires us to say only that U (is) false, since it (is) earlier than John's death, and that A-sentence-tokens possess only tenseless truth values. But I would emphasize that the unpalatable consequence Mellor draws from (18) is avoided by the conception of tensed truth expressed in (17); for (17) denies that the premature announcement U of John's death becomes true at a later time, when John dies, since (17) denies that U possesses truth at times when it is not present. Thus, the rejection of (18) does not force us to reject all notions of tensed truth. Indeed, some notion of tensed truth seems forced on us, given that Mellor's attempt to explain A-sentence-tokens solely in terms of the notions of tenseless truth and tenseless token-reflexive truth conditions failed, for reasons presented in chapter 3.

Graham Priest makes an interestingly different response to Mellor's argument:

Mellor finds it "absurd to say . . . that John's death posthumously verifies every premature announcement of it" (p. 170). But this is not absurd at all. What makes such an announcement premature is that it *was* false when it was made (though it *would be* true). Nothing that happens changes that. The utterance is verified in the sense of being made true (now), not in the sense of somehow showing the utterer to have been right then.[4]

Priest seems to be implying that Mellor's conception of nontenseless truth values includes only present tensed truth values and omits past and future tensed truth values, reference to which solves the problem that Mellor mentioned. It is not simply the case that the premature announcement U is now true, Priest suggests, but it is also the case that U has a past tensed truth value, namely, it was false. It is now the case that U, which is past, *is now* true but *was* false. Thus, Priest's complete multitensed truth conditions of U would seem to be

> (19) The token U of "John is dead" was true if and only if John was dead, is true if and only if John is dead, and will be true if and only if John will be dead.

The phrase "John will be dead" is ambiguous and is to be understood as meaning "either John is now alive but will be dead or John is now dead and will continue to be dead." There is also a further ambiguity. Suppose John was dead ten years ago but alive eleven years ago. Was U then true 11 years ago? Obviously not. Thus, (19) must be qualified to read that U was true at all and only the past times (sets of events/moments) to which John's state of being dead belonged and U will be true at all and only the future times to which this state will belong.

In order to decide between Priest's non-token-reflexive tensed truth conditions and the token-reflexive conditions I have presented, it is first necessary to further clarify my token-reflexive account. My account of tensed truth is not meant to imply that U is true if and only if U and John's death are both present *and false if U and John's death are not both present*. If U occurred ten years ago when John was dead, then U and John's death are not now both present, but it hardly follows from this that U is now false. According to my conception of tensed truth values, the "falsity conditions" of U are instead,

> (20) The token U of "John is dead" is false if and only if U is present and John's death is not present,

such that U's present tensed falsity, like its present tensed truth, requires U's presentness. If U is not present, it possesses no present tensed truth value but only a past or future tensed truth value (depending on whether U is past or future). But for Priest, if U is not present, it possesses a present tensed truth value and a past tensed truth value and a future tensed truth value. For me, a premature announcement of John's death was false when it occurred but is not possessing and will not possess any truth value in the present and future, when it is not occurring. U's *possession* of a truth value is a state or event (where a state or event is understood as the possession of an n-adic property by something), and this state of U is temporally located at all and only the times at which U is located and does not "continue to endure" after U itself has passed away. But there are truths about U's truth value that are located at times at which U is not located. In the present, it *is* (present tensed) true that U *was* false; and it *will be* true that U *was* false; but (to repeat) it *is not* true that U *is* true or false and it *will not be* true that U *will be* (at the future time) true or false.

There are two reasons to prefer this token-reflexive account to Priest's, one of which I will mention here, one at the end of the next section. My account, but not Priest's, conforms to the basic conventional rule of usage that sentence-tokens are used to express speaker's (writers, etc.) intentions. This expressive purpose is best facilitated if sentence-tokens obey the rule that their correspondences/

noncorrespondences to states of affairs obtain only as and when it is nondefectively intended to obtain by the speaker.

Let me explain. Sentence-tokens are semantically expressive (they express propositions) and psychologically expressive (they express the speaker's intentions), and it is the latter that I am concerned with here. The notions of speaker's intention and psychological expression are complex notions and involve a host of philosophical problems, but it is not necessary to explore these complexities and problems here in order to make the following simple point. Suppose Alice is now uttering U. Associated with her utterance are a number of speaker's intentions, such as the Gricean communication intention. But consider one of these intentions, her *alethic intention*. If this alethic intention is nondefective, it is an intention that U corresponds now to a state of affairs that obtains now, namely, the presentness of John's state of being dead. (Defective intentions are those that violate rules of use, such as an intention that U correspond now to John's present state of being heavy.) Alice has no (nondefective) intention about the correspondences/noncorrespondences that her token U will possess at some future time when U no longer is present. She does not intend that U, once past, *will* correspond to a state of affairs that *will* obtain but merely that it *does* correspond to a state of affairs that *does* obtain and (implicitly) that U *will have* corresponded to a state of affairs that *obtained* when U was present. Priest's theory requires us to suppose that U's correspondences/noncorrespondences have a "life of their own," so to speak—that only some of U's correspondences express speaker's alethic intentions and others obtain independently of what the speaker intended. Priest's theory requires us to suppose that U takes on a correspondence to John's death that Alice did not intend it to take on, namely, a correspondence that has presentness long after U (and perhaps even Alice herself) has passed away. Although I do not find Priest's thesis that *U possesses present tensed truth values when it is passed away* logically self-contradictory, it is inconsistent with what seems to me to be a basic rule of usage, *that semantic correspondences express speaker's intentions about semantic correspondences;* for Priest's theory entails that some semantic correspondences are not expressive of speaker's intentions. This is not a "knockdown argument" against Priest's theory; but it, along with the second reason I will present at the end of the next section, seems to render Priest's theory less attractive than the one I am advancing.

4.4 The Tensed Explanation of the Tenseless Token-Independent Rules Governing the Propositions Expressed by A-Sentence-Tokens

In the last section I provided an explanation of the tenseless rules of A-sentence-tokens by deducing these rules from the tensed rules of A-sentence-tokens. In this section, I shall offer an explanation of the tenseless rules governing the propositions expressed by A-sentence-tokens; these tenseless rules will be deduced from the tensed rules of these propositions.

I begin with a discussion of the *truth conditions* of A-propositions. The tensed truth conditions of these propositions are analogous to the tensed truth conditions of

A-sentence-tokens; for example, a present tensed A-proposition is presently true if and only if the proposition and the event it is about are both present. (In chapter 6 I shall argue at length against the traditional theory that propositions exist "outside of time"; for me, propositions and other abstract objects possess A-properties, such that there is nothing that is timeless.) The proposition *that the storm is approaching* is presently true if and only if *that the storm is approaching* and the storm's approach are both present. The analogy between the truth conditions of A-propositions and their expressing sentence-tokens is not accidental, since the latter truth conditions are based on the former. The tensed truth conditions of the token T_1 of "The storm is approaching" are deducible from premises stating the tensed truth conditions of *that the storm is approaching* and the premise that this proposition is expressed by T_1:

(1) The storm's approach is present.

(2) *That the storm is approaching* is present.

(3) *That the storm is approaching* is presently being expressed by T_1.

Therefore,

(4) T_1 and the storm's approach are both present.

Item (4) is the tensed truth condition of T_1. It is arguments such as (1)–(4) that provide foundations for the tensed rules of A-sentence-tokens that were discussed in the last section.

The tensed truth conditions of A-propositions enable the tenseless token-independent truth conditions of A-propositions discussed in section 3.2 to be explained. I there examined the sentence

(5) It was true that the era devoid of linguistic utterances is present

and specified its tenseless truth conditions as

(6) There (is) an era devoid of utterances earlier than the utterance of (5).

(7) There (is) some utterance-independent truth vehicle V that is expressed by the clause prefixed by "It was true that."

(8) V (is) true during the era devoid of utterances.

In line with our propositional theory, we can identify the truth vehicle V with the A-proposition *that the era devoid of linguistic utterances is present*. Since the tenseless truth of this proposition belongs to the truth conditions of the A-sentence (5), the tenseless truth conditions of this proposition indirectly belong to the tenseless truth conditions of (5). The proposition V's tenseless truth conditions, as specified in section 3.2, are that V (is) true when and only when V (is) simultaneous with the era devoid of utterances. More fully stated, these truth conditions are that

(9) The property of *being true* inheres in V when and only when the property of *being simultaneous with the era devoid of utterances* inheres in V.

These truth conditions concern the B-relation between the inherences of two properties in V; the inherence of the property of *being true* in V and the inherence in V of the property *being simultaneous with the era devoid of utterances* (call this era "E"). The locution "when and only when" is used in (9) instead of "if and only if" because it brings out more clearly that these two properties do—or at least can—tenselessly inhere in V *impermanently*, at only some of the B-times at which V exists. "When and only when" makes reference to dates, which is just what is needed; for impermanent properties tenselessly inhere in things (whether they be concrete things like people or abstract things like propositions) *only relatively to dates*. If P is an impermanent property of a thing T, then T's tenseless possession of P is not expressed by saying "T (is) P" but by saying "T (is) P at date D." The latter sentence implies that the *inherence* of P in T occupies the moment M or is a member of the set of events S that comprises the date D. Let us suppose that dates are sets of events. If events are regarded as inherences of properties in things (as I regard them), then the set of events D will be a set of simultaneous *property inherences*. One example of a date is a set of events composed of the *inherence* of walking in John, the simultaneous *inherence* of thundering in the atmosphere, the *inherence* of burning in a match that is simultaneous with the first two property inherences, and all other property inherences that are simultaneous with these three. To say that a thing T possesses a property P at a date D is to say that the inherence of P in T is a member of the set of simultaneous property inherences that constitutes date D.

It is these notions that flesh out the "when and only when" locution that belongs to the statement of V's tenseless truth conditions. The property of *being simultaneous with E* is an impermanent property of V, since at some of the B-times at which V exists, V is simultaneous with E, but at other of these B-times, V is not simultaneous with E. V exists at the B-time 10 billion years B.C. and at this time is simultaneous with E; but V also exists in 1986, and at this B-time it is not simultaneous with E. The assertion that V is true when and only when V is simultaneous with E means that the inherence of truth in V is a member of all and only the sets of simultaneous events of which the inherence of simultaneity with E is a member.

These tenseless truth conditions of the proposition V are *token-independent* in the sense that the existence of a sentence-token that expresses V is neither necessary nor sufficient to make V true. In fact, the existence of such a token is sufficient to make V false; for if V is expressed at a date D, it is false at D. These truth conditions are also mind-independent (in the sense discussed in section 3.6), for the existence of minds is neither necessary nor sufficient to make V true at any date.

These tenseless truth conditions are deducible from V's tensed truth conditions. V is presently true if and only if V is present and E is present. The entailment is straightforward:

(10) V is true (in the present tensed sense) if and only if V and E are both present.

Therefore

(11) V (is) true when and only when V (is) simultaneous with E.

Item (11) may be phrased in the "if and only if" idiom if we add a date-expression:

(12) V (is) true at a date D if and only if V (is) simultaneous with E at D.

Since the clause stating the tensed truth conditions, "V and E are both present," entails, but is not entailed by, the clause stating the tenseless truth conditions, "V (is) simultaneous with E at (a date) D," it may be said that the tenseless truth conditions of V are reducible to V's tensed truth conditions.

The token-independent *confirmation conditions* of the propositions expressed by A-sentences were discussed in section 3.3. I there examined the present tensed A-sentence

(13) I am not uttering anything

and indicated that the proposition p that I did not express, but might have expressed, at some date D is confirmed as true at D by observations that I am not uttering anything at D. The tenseless observation sentence that reports the observation confirming p to be true at D,

(14) It (is) observed that I (am) not uttering anything at D

is entailed by the tensed observation sentence

(15) It is observed that my not uttering anything and D are both present;

but (14) does not entail (15).

In section 3.4 I presented evidence that the rules of usage of A-sentences that are followed in ordinary and scientific language imply that normal A-sentences (A-sentences that do not mention their expressing sentence-tokens) express propositions that do not refer to sentence-tokens that express them. According to these rules, each token of a normal A-sentence about an event E entails

(16) There (is) something sententially expressed about E whose truth value is logically and empirically independent of its being sententially expressed.

It is easy to see how (16) is entailed by the assertion that A-sentence-tokens express A-propositions, for A-propositions *attribute A-properties to events* and consequently are dependent for their truth value only upon the A-properties possessed by themselves and the events they are about. Since whether or not the A-proposition is expressed is irrelevant to whether or not the event and the A-proposition possess the appropriate A-properties, the expression or nonexpression of the proposition is irrelevant to its truth value.

The thesis that A-sentences express A-propositions also accounts for the facts about the *logical identity* of A-sentences discussed in section 3.5. I there indicated that the theory that A-sentences express propositions that refer to their expressing tokens is unable to account for the (explicit) logical identity and equivalence of simultaneous utterances of synonymous A-sentences. Given the assumption that these sentences express A-propositions, the logical identity and equivalence of simultaneous utterances of

(17) Jane is leaving

and

(18) Jane is departing

is easily explained; for this assumption implies that these two utterances express one and the same A-proposition, a proposition that attributes presentness to Jane's leaving.

This characterization of the token-independent and proposition-involving rules of A-sentence-tokens enables me to present my second reason for preferring my account of tensed truth values and truth conditions to Graham Priest's. According to Priest, if Alice uttered a token U of (17) one week ago, when Jane was not leaving, then U was false, but U becomes true at a later time if Jane's leaving acquires presentness then. A defender of Priest may allege that this account is preferable to mine, since it explains how we ordinarily talk about tensed truth. For example, one might say "What Alice stated was false one week ago but is now true, since Jane is now leaving." But it seems to me that "what Alice stated" here refers to the proposition expressed by Alice's utterance U, not to U itself. The proposition *that Jane is leaving* was false one week ago, when Jane's leaving did not have present-ness, but is true now, when Jane's leaving does have presentness. There is an important difference between U and the proposition that enables the latter alone presently to have a present tensed, as well as a past tensed, truth value, namely, *that the proposition, but not U, both was present and still is present.* U was present one week ago but is now passed away, but the proposition continued to be present and for this reason continued to have a present tensed truth value. In my judgment, Priest's theory confuses the truth values and conditions of A-sentence-tokens with the values and conditions of the propositions they express, for the latter alone perdure in a way that enables them to have changing truth values over the relevant stretches of time. At the very least, Priest's theory of tensed truth fails to explain or reflect the temporal asymmetry between tokens and propositions, since his theory implies that presently existing propositions and the long past tokens that once expressed them *both* have present tensed truth values. Priest's theory fails to respect or even consider the following line of reasoning: "Jane is leaving; therefore, the proposition *that Jane is leaving* is now true, since this proposition is *now* simul-taneous with her leaving; but the long past utterance U is not now true, since it is not *now* simultaneous with her leaving." I believe that this consideration (i.e., the temporal asymmetry between propositions and sentence-tokens) along with the considerations about the expression of speaker's alethic intentions discussed in the last section, render Priest's theory less preferable than the one here advanced.

In this and the previous section, I have argued that the tenseless token-reflexive rules (discussed in chapter 2) and the tenseless token-independent rules (discussed in chapter 3) of A-sentence-tokens are both compatible with, and explained by, their tensed rules. Since these tensed rules, especially the token-independent ones, involve A-propositions, the arguments offered go a long way toward showing that the crucial step in the argument (8)–(13), namely, the premise that "besides B-propositions, the only other candidate for the propositions expressed by A-sentences are A-propositions" (see section 4.2) is true. Chapters 2 and 3 aimed to show that B-propositions are not viable candidates for these expressed propositions, and my arguments in the past two sections have endeavored to show that A-propositions *are* viable candidates. But these arguments are incomplete, for I have not yet accounted for other elements of the linguistic data gathered in Chapters 2 and 3. I have in mind the distinction between A-sentences containing A-indexicals ("now," "today," etc.) and A-sentences not containing them and the associated problem of the *cognitive significance* of the indexicals. The explanation of these data in terms of the hypothesis that A-sentences express A-propositions is the next task that confronts us. Once this is accomplished, I will proceed to defend the logical coherency of the A-proposition theory against arguments based on McTaggart's paradox and to develop the presentist theory of language and time.

4.5 The Tensed Explanation of A-Indexicals and their Relation to A-Nonindexicals

A-indexicals include "now," "at present," "currently," "formerly," "eventually," "today," "yesterday," "tomorrow," and other such locutions. They include these words both in their pronomial and adverbial uses. Examples are

   (1) She is leaving now.

   (2) Now is the time to leave.

In (1), "now" is an adverb, in (2), a pronoun. A-words that are nonindexical include A-copulae ("is," "was," "will be," and the like) and A-verbs ("ran," "screamed," etc.). A-adjectives ("present," "past," "future," etc.) are nonindexical if they are prefixed by an A-copulae (as in "the storm is present"); but they function indexically if they are not so prefixed (as in "The present storm is the most dangerous we've had in years").

The distinction between A-indexicals and A-nonindexicals played an important part in my studies in chapter 3. I there distinguished between

   (3) It was true that the era devoid of linguistic utterances is present

and

(4) It was true that the era devoid of linguistic utterances is occurring now

and used this distinction to argue that (3)—but not (4)—could be uttered with truth. They have *different* tenseless truth conditions. However, in chapter 2, I treated A-sentence-tokens containing the A-indexical "now," such as

(5) The era devoid of linguistic utterances is occurring now

as having the *same* tenseless truth conditions as A-sentence-tokens not containing A-indexicals, such as

(6) The era devoid of linguistic utterances is present.

I implied (5) and (6) are both true if and only if they occur simultaneously with the event or era they denote. Clearly, there is some interesting semantic difference between the semantic contents of "is occurring now" and "is present" that enables the *extensional* sentence-tokens (5) and (6) to have the same truth conditions and the *intensional* constructions (3) and (4) to have different truth conditions. No inkling of this difference in semantic content was given in the last two sections. Yet it is important that differences of this sort between A-indexicals and A-nonindexicals be explained, otherwise the hypothesis *that A-propositions are viable candidates for being the propositions expressed by A-sentences* would be left without adequate justification. This hypothesis is justified only if it can be shown to explain *all* the tenseless rules of usage of A-sentences and only some of these rules were explained in the last two sections, the token-reflexive and token-independent rules that are *common* to A-sentences with A-indexicals and A-sentences without them. There are also token-reflexive and token-independent rules unique to A-sentences with A-indexicals and other rules unique to A-sentences without A-indexicals. These different rules must be explained by a theory that also explains why these sentences share in common the rules discussed in the last two sections. This is no idle or easy task but is a central problem facing the A-theory, considering that the standard A-theory is unable to explain these differences and similarites among the rules of usage (as I shall argue shortly). I believe the difficulties can be overcome and the explanation achieved if the A-theory is integrated with the absolutist theory of time, the theory that times are not sets of events but event-independent moments. If this explanation is achieved, the main aim of chapters 1–4 will be accomplished, the demonstration that *only* the hypothesis of A-propositions is able to explain all the semantic properties of A-sentences. This demonstration is significant because it entails (given that some A-sentence-tokens are true) that A-properties are exemplified and thus that the tensed theory of time is true.

The task immediately before us is to gather all the linguistic data that are relevant to the A-explanation of the rules of usage specific to A-sentences with A-indexicals and the rules specific to A-sentences without such indexicals. These data are found in linguistic contexts that reveal the differences in behavior between A-sentences with A-indexicals and A-sentences without them. These are *temporal and modal contexts,* which involve sentences prefixed by temporal operators such as

"It was true that" or modal operators such as "It is possible that." The essential fact that needs to be explained, as we shall see, is that *A-indexicals, but not A-nonindexicals, are insensitive to temporal and modal operators.*

An expression is insensitive to temporal operators if and only if it refers to the same A-time when it is prefixed by a temporal operator as it refers to when it is not prefixed by a temporal operator. Its reference to an A-time is not affected by temporal operators. Let us take "today" as a sample A-indexical. If "today" is tokened several times in the same circumstance such that in one tokening it belongs to a sentence-token lacking a temporal operator, in another tokening to a sentence-token that has a past tensed temporal operator, and in a third to a sentence-token that has a future tensed temporal operator, then in each of these tokenings, "today" will refer to the same A-time. Compare the Thursday tokenings of

(7) Today is Thursday.

(8) It was true yesterday that today is Thursday.

(9) It will be true tomorrow that today is Thursday.

"Today," as tokened in each of these cases, has the same reference, namely, to the present day, the day of the tokening, Thursday. The operators do *not* determine "today" to refer to the day referred to by the operator—yesterday and tomorrow.

An A-expression is insensitive to *modal operators* if and only if it refers to the same A-time when it is within the scope of a modal operator as it refers to when it is not within the scope of such an operator. Compare Thursday tokens of

(7) Today is Thursday.

(10) It is logically possible that today is the last day of time.

(11) It is logically necessary that today is a day.

It is clear that the Thursday token of "today" in (7) refers to Thursday, but the same is true of the tokens of "today" in the tokens of (10) and (11). In the token of (10), "today" does not refer to just any day that has the property in some merely possible world W of being the last day of time; it refers to *today*—the day I actually utter (7)—and asserts that this day has in W the property of being the last day of time. Likewise, in uttering (11), what I am saying is that *today,* the day I am uttering (11) in the actual world, has the property of *being a day* in each possible world in which this day exists.

The task of explaining these data regarding temporal and modal contexts in terms of the A-theory is important, since previous attempts (few that they have been) have failed. The main attempt is exemplified by the recent approaches of Pavel Tichy and George Schlesinger,[5] who assert that words such as "now" or "today" indirectly refer to times by expressing the propositional definite description (the sense) *whatever time is present.* In the case of "today," this word allegedly expresses *whatever day is present.* However, this standard A-theory of temporal indexicals is ruled out by the insensitivity of these indexicals to modal operators. Compare Thursday tokens of

(12) There is some merely possible world W in which today is the last day of time

and

(13) There is some merely possible world W in which whatever day is present is the last day of time.

Clearly, the "today" in the token of (12) refers to the day on which it is actually uttered—Thursday, May 1, 1986. But "whatever day is present" in the token of (13) does not convey a transworld identity between the day that is present in the actual world and the day that is present in W. What satisfies the description *whatever day is present* in W need not be what satisfies this description in the actual world. In the sentence

(14) Whatever day is present is the last day of time,

the description does refer to the present day in the actual world, but the modal operator in (13) acts upon this description and dissolves its tie to the day that is actually present. Hence, all we learn from (13) is that *some* day is present in W and that that day is the last day of time.

A second A-theory of the indirect reference of "today" is able to solve this problem by holding that "today" expresses the modally qualified definite description *whatever day is actually present*. This theory, although not propounded by anyone, is suggested by Plantinga's theory of proper names.[6] Plantinga believes that the rigidity of proper names in modal contexts can be explained if we assume that they express modally stable definite descriptions—descriptions that have the same referent in every possible world to which they refer. "The person who invented bifocals" is modally unstable, since it refers to different people in different worlds. In the actual world, Benjamin Franklin invented bifocals; but in world W Jane Doe invented them. By contrast, the definite description "the person who actually invented bifocals" refers to the same person in every world (to which the description refers), namely, to Benjamin Franklin. In a corresponding fashion, if we add "actually" to "whatever day is present" we obtain a modally stable description that simulates the behavior of "today" in modal contexts. Suppose I utter on Thursday, May 1, 1986,

(15) There is some merely possible world W in which whatever day is actually present is the last day of time.

The utterance of (15) does refer to the day that is May 1, 1986 in the actual world and asserts that in W this day is the last day of time. The adverb "actually" renders the description "whatever day is actually present" no less insensitive to modal operators than is "today." So it seems that "whatever day is actually present' does translate "today."

But this A-theory founders on the reef of temporal operators, for "today"—but not "whatever day is actually present"—is insensitive to temporal operators. Compare Thursday utterances of

(16) It was true yesterday that today is Thursday

and

(17) It was true yesterday that whatever day is actually present is Thursday.

Item (16) is true, but (17) is false. The modally qualified definite description in (17) is sensitive to the operator and is determined by the operator to lose its reference to the day of utterance and to refer, instead, to the day referred to by the operator—to the day that is actually present *yesterday,* to Wednesday. Item (17) asserts that the referent that the descriptive sense *whatever day is actually present* possessed *yesterday* is Thursday; this assertion is false because the referent that this sense possessed yesterday is Wednesday. Item (16) on the other hand, asserts that it was true yesterday that the referent possessed *today* by "today" is Thursday—which is true. The indexical "today" is not made to lose its reference to the day of its utterance by the operator. Consequently, (16) is not translated by (17).

These results suggest that A-indexicals such as "today" do not express propositional definite descriptions—senses—but are, instead, directly referential. This leads us to the third A-theory of "today" that I shall consider before I present the one I believe to be correct. The third A-theory (which I developed—but did not evaluate or endorse—in "Problems with the New Tenseless Theory of Time")[7] entails that each token of "today" refer directly to the daylong set of events of which that token is a member. These sets of events are "dates" in the sense defined in section 2.1. But this is not all there is to this theory of "today," given that it is an A-theory. Each token of "today" also ascribes the property of presentness to its direct referent. The token of "today" that is a member of the day-long set of events, May 1, 1986, directly refers to this set and ascribes presentness to it. The relevant token of "Today is Thursday" would be translated by

(18) [May 1, 1986], which is present, is Thursday,

where "[May 1, 1986]" directly refers to the set of all and only those events that are, in fact, 1,985 years and four months later than the birth of Christ. This third theory seems promising at first glance; for the directly referential description "[May 1, 1986], which is present" is similar to "today" in being insensitive to *both* modal and temporal operators. If "[May 1, 1986], which is present" is placed after any modal or temporal operator, its reference will remain unchanged: it will still refer to the day that is actually May 1, 1986. For example, "There is some possible world in which [May 1, 1986], which is present, is the last day of time" refers to the same day-long set of events in both the actual and possible world and thus is similar to the corresponding tokens of the "today"-sentences (10) and (12) in being rigid. And clearly, an utterance on Thursday, May 1, of "It was true yesterday that [May 1, 1986], which is present, is Thursday" refers to the day of its utterance, Thursday, and consequently has the same truth value as a Thursday token of "It was true yesterday that today is Thursday."

The problem with this third theory is that it is unable to explain the behavior of

A-indexicals in *all* modal contexts. There are some counterfactual contexts in which it produces counterintuitive results. Let us compare May 1, 1986 utterances of

(19) World War III did not begin today, but it might have

and

(20) World War III did not begin on [May 1, 1986], which is present; but it might have.

Item (19) entails

(21) There is some possible world W in which the start of World War III occurs today.

Item (20) entails

(22) There is some possible world W in which [May 1, 1986] includes the start of World War III.

The referent of "[May 1, 1986]" is a set of events, the day-length set of events that actually has the property of being 1,985 years and eighty-nine days later than Christ's birth. If this set does not actually include the start of World War III, it is logically impossible that it include it; for *sets contain their members essentially.* Sets contain all and only the members they actually contain in every possible world in which they exist. Thus, (20) is self-contradictory: it asserts that the set [May 1, 1986] contains a member in a merely possible world W that it does not actually contain. But (19) is not self-contradictory; indeed, (19) is true. This shows that "today" does not directly refer to the day-long set of events that includes the event of its tokening.

Nevertheless, I believe that a sound theory of A-indexicals can be constructed that is formally analogous to this directly referential theory. This fourth theory retains the assumption that tokens of A-indexicals are directly referential but adds to it the further assumption that the absolutist theory of time is true. The absolutist theory holds that times are not sets of events but *moments* that are accidentally occupied by events (if occupied at all). The moment M that is actually occupied by all and only the events that comprise the set [May 1, 1986] is occupied by different events in different worlds and in some worlds (those in which there is "empty time") is not occupied by any events at all. On this theory, the May 1, 1986 token of "today" directly refers to the moment M and refers to this moment in each world to which the token refers. The token of (19) that occurs on this day is translated by

(23) World War III did not begin at M, which is present; but it might have.

Item (23) is clearly consistent, since M is not *actually* occupied by the start of World War III but is occupied by this event in some other possible world.

Before I show how this A-theory completely explains the behavior of A-indexicals, I should note that its assumption of the absolutist theory of time need not be taken as a drawback to it. There are independent reasons to accept the absolutist theory, as I have argued in *The Felt Meanings of the World*.[8] A further defense of this theory is offered in "Kant and the Beginning of the World";[9] and a number of other authors (notably Sydney Shoemaker, Richard Swinburne, and W. H. Newton-Smith), have provided good arguments for it.[10] Thus, I believe that the assumption of the absolutist theory of time is an advantage, rather than a weakness of this fourth theory of A-indexicals, and provides external justification for it.[11]

I indicated how this absolutist A-theory explains the modal sentence (19). What is needed next is an explanation of the modal sentence (10) and the temporal sentence (16). The sentence

(10) It is logically possible that today is the last day of time

as uttered on May 1, 1986 expresses the same proposition as

(24) There is some possible world in which M, which is present, is the last day of time

The tokens of (10) and (24) both assert that there is some possible world W that is just like the actual world except for the fact that M (the day-long moment on which [10] is actually tokened) has the property of being the last day of time. In W, the moment M includes the events that it actually includes (or, at least, all those events that are consistent with M's being the last day of time) and differs merely in its temporal relations to other moments: there are no later moments than it.

The explanation of the temporal sentence (16) is more complicated. The Thursday token of

(16) It was true yesterday that today is Thursday

directly refers to the day of its utterance, M; ascribes presentness to M; and asserts that it was true yesterday that M is Thursday. Manifestly, the sentence "It was true yesterday that M, which is present, is Thursday" does not assert that the day that was present yesterday (call it M_0) is Thursday but that *M* is Thursday.

Further explanation is needed, however, considering that the Thursday token of "today" ascribes presentness to M, yet M is *not* present on Wednesday. Undeniably, "M, which is present, is Thursday" is not true on Wednesday. This explanation is best achieved if we compare a Thursday token of (16) with a Thursday token of

(25) It was true yesterday that Thursday is present.

Item (16) includes the A-indexical "today" in its embedded sentence ("Today is Thursday"); and (25) includes, instead, the A-nonindexical "is present" in its embedded sentence ("Thursday is present"). Now it is a linguistic datum that the Thursday token of (16) is true and the Thursday token of (25) is false. What explains this? I believe the explanation is that the operator "It was true yesterday that" behaves differently in (16) and (25). Although operators act upon the propositions expressed by the embedded sentences, rather than upon the embedded sentences themselves, their behavior is sometimes most easily studied if it is provisionally assumed that they act upon the embedded sentences themselves. This assumption provides illumination in cases where the behavior of the operator in relation to the embedded sentence is similar to its behavior toward the expressed propositions. If we adopt this assumption for a moment, we will see that the operator must be taken as acting upon "Thursday is present" in a different way from that in which it acts upon "today is Thursday." It acts *directly* on the former sentence and *indirectly* on the latter. It acts directly on "Thursday is present" in that it asserts this very sentence to possess the property of truth yesterday, when it was Wednesday. This explains why the Thursday token of (25) is false; for clearly, the sentence "Thursday is present" is false on Wednesday. But we cannot assume that the operator also acts directly upon "today is Thursday" in (16). If it did, the Thursday token of (16) would also be false; for manifestly, the sentence "today is Thursday" was not true yesterday, on Wednesday. Given that the Thursday token of (16) is true, we must assume that the operator acts *indirectly* upon the embedded sentence. It asserts that some other sentence was true yesterday, some sentence whose being true yesterday entails that the embedded sentence is true today. Such a sentence is "Tomorrow is Thursday"; this sentence was true yesterday only if "today is Thursday" is true today.

If we take these sentential analyses of (16) and (25) as a key to their propositional analyses, we shall arrive at the idea that the proposition expressed by the embedded sentence-token "Thursday is present" is acted upon directly by the operator and that the proposition expressed by the sentence-token "today is Thursday" is acted upon indirectly by the operator. These analyses correspond exactly to the theory of A-propositions I am proposing. According to this theory, the token "Thursday is present" expresses the A-proposition *that Thursday is present*. This proposition, although true on Thursday, is false on Wednesday. If we assume that the operator acts directly upon this proposition, then we shall take the Thursday token of (25) as asserting that this proposition was true one day ago, on Wednesday. Since this proposition was false then, this token is false. Thus, we have a propositional explanation of the falsity of the Thursday token of (1) that mirrors the sentential explanation.

A similar correspondence holds between the propositional and sentential analysis of the Thursday token of (16). According to my account of "today," the Thursday token of "Today is Thursday" expresses the proposition *M, which is present, is Thursday*. This proposition is also false on Wednesday, since on Wednesday, M is not present. Thus, to account for the truth of the Thursday token of (16), we must assume that the operator acts indirectly upon this proposition. The operator must be taken as asserting that some other proposition was true yesterday, some proposition whose being true yesterday entails that the proposition *M, which is present, is Thursday* is true today. Such a proposition is *M, which is future by one day, is Thursday*. This proposition is the one expressed on Wednesday by "Tomorrow is Thursday."

This discussion indicates that—and how—the absolutist A-theory explains the behavior of "today" in modal and temporal contexts. An explanation of this sort is possible for the modal and temporal rules of use of all present tensed A-indexicals. The tenseless rules of use unique to present tensed A-indexicals may be summarized by the statement that uses of these indexicals refer to the time of their use even in temporal and modal contexts. This general rule is explained by the A-theory that uses of these indexicals directly refer to the moment of their use and ascribe presentness to this moment. Indeed, the preceding pages may be taken as a demonstration that the absolutist A-theory entails this general tenseless rule.

Some of the ideas implicit in the preceding pages also provide an A-explanation of the tenseless rules of usage of A-nonindexicals. These rules are summarized by the statement that uses of A-nonindexicals convey that a certain temporal property is possessed at the time of their tokening if and only if these uses are not operated on by the relevant temporal or modal operators. If they are operated upon, they convey that the temporal property is possessed at the time to which the operator refers. The main example of an A-nonindexical used in the preceding pages is the predicate phrase "is present" appearing in "whatever day is present" and "Thursday is present." The foregoing discussions suggest that uses of such A-nonindexicals do not directly or indirectly refer to the time of their use but *merely ascribe an A-property.* Whereas the semantic content of "today" (as pronomially used on a certain occasion) can be represented as

(26) M, which possesses presentness,

the semantic content of "is present" (or simply the present tense of "is") can be represented as

(27) ( ) possesses presentness.

Since (27) does not include, or refer to, the moment of its use, it is able to convey (in modal or temporal contexts) that some item *at some other time* has presentness.

These notions can be given a concrete application in regard to the sentences "Thursday is present" and "Today is Thursday." If we assume that "Thursday" is used to express a descriptive sense (e.g., *whatever day is the fourth day of the week*), then we can represent the semantic content of "Thursday is present" by

(28) ⟨The sense of "Thursday", presentness⟩.

Item (28) ascribes the property of presentness to the referent of the sense of "Thursday." Note that the sentence "Thursday is present" expresses the proposition (28) on each occasion of its tokening. This proposition changes its truth value over time; for each week, it is false from Friday to Wednesday but true on Thursday.

By contrast, the A-indexical sentence "Today is Thursday" expresses a different proposition on each different day it is tokened. If it is uttered on Wednesday, April 31, 1986, it directly refers to the moment M_0 that contains all and only the events that actually occur on this date. If it is uttered on Thursday, May 1, 1986, it directly refers to the different moment M, which is actually occupied by different events. The Wednesday utterance expresses the proposition

(29) $\langle\langle$The sense of "Thursday," $\langle M_0$, presentness$\rangle\rangle$, being identical$\rangle$,

the Thursday utterance, the proposition

(30) $\langle\langle$The sense of "Thursday," $\langle M$, presentness$\rangle\rangle$, being identical$\rangle$.

Item (29) ascribes presentness to the moment M_0 (which is a constituent of the proposition) and asserts that the relation of identity holds between M_0 and the referent of the sense of "Thursday." The same holds, mutatis mutandis, of (30). Item (29) is false on every day; and (30) is false on every day but Thursday, May 1, 1986, when it is true.

This explains why

(3) It was true that the era devoid of linguistic utterances is present

and

(4) It was true that the era devoid of linguistic utterances is occurring now

have different truth conditions. The operand in (3), the clause "the era devoid of linguistic utterances is present," expresses the same proposition on each occasion of its use, that is,

(31) $\langle$The sense of "the era devoid of linguistic utterances," presentness$\rangle$.

This proposition *is* false but *was* true when the era devoid of linguistic utterances possessed presentness. But the operand in (4) expresses a different proposition on each occasion of use. As used on May 1, 1986, it expresses the proposition

(32) $\langle\langle$The sense of "the era devoid of linguistic utterances," $\langle M$, presentness$\rangle\rangle$, being identical$\rangle$.

Item (32) asserts that M has presentness and that M is identical with the referent of the sense of "the era devoid of linguistic utterances." Manifestly, (32) is true at no time, past or present, given that May 1, 1986 contains numerous linguistic utterances.

The A-theory that I have been developing also explains why A-sentences with A-indexicals share rules of use in common with A-sentences that lack such indexicals. These common rules are exemplified by the common truth conditions of such extensional sentences as

(33) E is present

and

(34) E is occurring now,

which directly refer to the same event E. Each of these sentences is true if and only if its tokening is present when E is present. But how can this be explained, given that they express different propositions? On each occasion of its tokening, (33) expresses the proposition

(35) $\langle$E, presentness$\rangle$.

Item (34) expresses a different proposition at each different time it is tokened; on May 1, 1986, it expresses

(36) $\langle\langle$E, $\langle$M, presentness$\rangle\rangle$, occupying$\rangle$,

which states that M is present and is occupied by E. In section 4.4 I indicated that propositions such as those expressed by uses of (33) and (34) have the same tensed truth conditions, namely, that they are presently true if and only if they and the events they are about are both present. It might be objected that this truth condition applies to the proposition (35) but not to the proposition (36), since (36) involves the moment M and consequently that the presentness *merely* of the proposition (36) and the event E are insufficient to make the proposition true. An extra and logically independent truth condition must be added, *that M is present*. It follows, the objection continues, that my A-theory fails to explain or to cohere with the fact that uses of (33) and (34) express propositions with the same truth conditions.

This objection is unsound, since M is a part of the proposition (36) and is occupied by E and thus is present if the proposition and E are present. Hence, the A-theory developed in this section is able to explain the truth-condition similarity of propositions such as (35) and (36).

This theory is also able to explain the similarity of the truth conditions of the sentence-tokens that express these propositions. This explanation, however, is considerably different from the one just given. In section 4.3, I implied that any token T of "E is present" is presently true if and only if S and E are both present and that any token S of "E is occurring now" is presently true if and only if S and E are likewise copresent. It may be obvious that my A-nonindexical theory entails T has these truth conditions, but it might be doubted that my A-indexical theory allows S to possess them. S refers to a moment and requires the presentness of the moment; but its presentness is not entailed by S's presentness, since the moment is not a part of S. It seems that an additional truth condition is needed, namely, *that the moment is present*.

This is not the case. S exists in many possible worlds and in some of these worlds expresses propositions other than the one it actually expresses (which I shall assume is [36]). Suppose that S is a vocal utterance; in that case, it is a physical event, an air vibration that possesses a certain amplitude, intensity, frequency, and wavelength. If S is *that* air vibration ("*dthat*" air vibration in Kaplan's sense), then these wave properties are necessary to S. If S is a low-pitched whisper in the actual world, it is counterintuitive to suppose that S—*that* very air vibration—is a high-pitched scream in some possible world W. One might also suppose with some plausibility that S is necessarily an utterance, a sentence-token, that is, that it necessarily exemplifies the properties of *being a token of (34)* and *being uttered by somebody.* Perhaps *being uttered by P,* where P is the person who actually uttered S, is also a necessary property of S. Now it is intuitively plausible to think that there is some possible world W at which P utters S but at a different time than the time at which P actually utters S. Indeed, there is "intuitive content" (to use Kripke's phrase) to the notion that S might have been uttered by P at a different time than P actually uttered it. P might have whispered it a moment later, in which case, the copresence of S and the moment M is not a necessary condition of S's truth. S refers to different moments in different worlds; consequently, the presentness of no one of these moments is necessary to S. What is necessary to S's truth—what belongs to each logically possible world in which S is true—is only the copresence of S and E. If S and E are both present, then S is true, regardless of which moment is present. Thus, S does not have different truth conditions from T; and my theory of A-indexicals is consistent with the semantic similarity between S and T.

This concludes the main brunt of my argument that the A-theory developed in this section explains both the differences and similarities between A-sentences with A-indexicals and A-sentences without them. There remain two further important sets of data about A-expressions that need to be explained by the tensed theory of time. I argued in chapter 2 that the data pertaining to the *quasi indicators* that represent uses of A-expressions and the data pertaining to the *cognitive significance* of A-expressions was not satisfactorily explained by the tenseless theory of time. A tensed explanation of these two sets of data is provided in sections 4.6 and 4.7, respectively.

4.6 The Tensed Theory of Temporal Quasi Indicators

The tensed explanation of A-indexicals and A-nonindexicals offered in the last section enables us to formulate a theory of *temporal quasi indicators.* Quasi indicators, we recall from section 2.5, are mechanisms, discovered by Castañeda, that serve to express the semantic content of someone's uses of indexicals (indicators). If Alice utters at noon "David is now asleep," then her use of "now" is (according to Castañeda) represented by "then" as embedded in the oratio obliqua clause of

(37) At noon, Alice (knows) that David (is) asleep then.

The embedded clause "David (is) asleep then" allegedly expresses the same proposition expressed by Alice's noon utterance of "David is now asleep." However, as I argued in section 2.5, such allegations are false, since Alice's utterance—but neither (37) nor its embedded clause—entails "David's sleep is present, rather than past or future."

I propose, in this section, to introduce a category of *tensed quasi indicators* that are capable of expressing the semantic content of someone's use of A-indexicals such as "now" and "today." I believe a noon utterance of the following sentence contains a quasi indicator that expresses the semantic content of Alice's noon token of "now":

(38) It is now noon, and Alice now knows that David is now asleep.

I believe that the third occurrence of "now" functions both as an indicator (indexical) and as a quasi indicator. It purports to convey, and succeeds in conveying, the speaker's indexical reference to noon *and* purports to convey, and succeeds in conveying, Alice's indexical reference to noon. The third occurrence of "now" achieves this by expressing the semantic content (⟨*noon, presentness*⟩) dictated by the rules of use of "now" as an indexical *and* by purporting (by virtue of being operated on by the oratio obliqua operator) that this semantic content is also the semantic content of some A-indexical token produced (physically or mentally) by Alice. By virtue of this, the noon utterance of the embedded clause in (38), "David is now asleep," purports to express and succeeds in expressing the proposition also expressed by Alice's noon token of this locution.

These remarks contradict Castañeda's fundamental principle that "an indicator *i* . . . expresses no attribution of reference or mechanism of reference to anyone— even if *i* appears in *oratio obliqua* clause (i.e. clause in indirect speech."[12] This implies that indicators appearing in oratio obliqua do not attribute to anyone or purport to represent anyone's uses of these indicators themselves. Thus, Castañeda writes of "Mary believes that I am here now" that "Mary is not being attributed the references expressed by 'I', 'here' and 'now'. . . . These indicators all express speaker's reference only."[13] I admit that some uses of indicators in oratio obliqua clauses express speaker's references only; but I believe this is false as a universal principle, since the third occurrence of the indicator "now" in (38) purports to convey both the speaker's indexical reference to noon and Alice's indexical reference to noon. The fact that it purports to convey Alice's indexical reference is demonstrated as follows. If the noon utterance of (38)—call it N—did not purport to express the semantic content of an indexical reference to noon made by Alice, then N would not have the truth conditions it in fact possesses; for N is true *only if Alice refers to noon via a present tensed A-indexical such as "now."* If Alice does not refer to noon indexically and either does not refer to noon at all or refers to it merely through some nonindexical description or name, then N is false. For example, if Alice knew at noon merely what is expressible by "David (is) asleep at whatever time (is) noon, June 19, 1989" or "David (is) asleep at T" (where "T" is a name of noon), she would not know what is expressible at noon by "David is now asleep," and N would thereby be false. Utterance N entails that Alice refers to noon indexically and thereby that Castañeda's general principle is false.

But this analysis of the use of "now" in (38) as a quasi indicator is too strict. Suppose that Alice tokens at noon the A-sentence "David is asleep" but not the A-sentence "David is now asleep." The former sentence contains no indexical. Is the utterance of (38) then false? I think our linguistic intuitions would lead us to say that the utterance of (38) remains true. The proposition Alice knows by virtue of tokening at noon "David is asleep" obviously (to all ordinary language users) entails at noon the proposition expressible at noon by "David is now asleep"; thus, Alice knows at noon something that obviously entails the proposition expressed by the embedded clause in (38). This suggests that a less strict and more plausible theory of the use of "now" as a quasi indicator in (38) is that (38) purports to express an A-indexical proposition known by Alice *or obviously entailed by* some nonindexical A-proposition known by Alice.

If we introduce "now" (in some of its uses in indirect speech) as a tensed quasi indicator, then some of the properties Castañeda uses to define quasi indicators must be rejected. Castañeda believes it to be essential to words that appear in sentences as quasi indicators that these words do not also express indexical references made by the speaker.[14] I believe, however, that this property is not definitive of quasi indicators, since the third occurrence of "now" is a quasi indicator and expresses an indexical reference made by the speaker. A second allegedly essential property of quasi indicators that must be rejected is that they are means of making somebody's "indexical reference both interpersonal and enduring, yet preserving it intact."[15] This does not apply, since the quasi indicator "now" in the sentence (38) does not convey Alice's reference in an enduring and repeatable manner; it convey's Alice's reference *only if uttered at noon*. If uttered at a later time, "now" refers to that later time and thereby no longer preserves intact Alice's noon reference. If uttered at 1:00, it expresses the semantic content

(39) ⟨1:00, presentness⟩,

whereas the semantic content of Alice's noon utterance of "now" is

(40) ⟨Noon, presentness⟩.

A third characteristic that must be rejected is that quasi indicators require an antecedent to which they refer back. Castañeda's theory is that "then" essentially refers to some temporal antecedent, such as "At noon" (as in [37]). But the third occurrence of "now" in (38) does not require the antecedent "It is now noon." The elimination of this antecedent eliminates information about the B-time of Alice's knowledge and David's sleeping, but it does not deprive the last "now" in (38) of its quasi indicator function. The sentence "Alice now knows that David is now sleeping" is true only if Alice now knows of David's sleeping in terms of a sentence that contains, in the right way, a present tensed A-expression.

Somebody's present use of an A-indexical can be represented by a quasi indicator; but somebody's past or future use of an A-indexical can be presently represented only by an *indicator–ascriber*. An indicator–ascriber does not express the semantic content of somebody's use of an indexical but merely ascribes to somebody the use of an indexical of a certain sort. A sentence-token containing a past tensed indicator–ascriber expresses the proposition *that somebody grasped a proposition that she expressed by a mental or physical sentence-token containing a certain sort of indexical expression.* Suppose it is now 6:00, June 19, 1989 and that I know that David is not now sleeping. I wish to convey that Alice knew at noon, June 19, 1989 what she then expressed by "David is now sleeping." I may utter

> (41) When noon, June 19, 1989 was present, Alice knew of this time that it was present and that it contained David's sleeping.

The expression "Alice knew of this time that it was present" is meant to convey that Alice's knowledge of the presentness of noon, June 19, 1989 was *de re,* not *de dicto.* Alice did not know (or may not have known) at noon *that noon June 19, 1989 possessed presentness;* but she did know at noon *of noon, June 19, 1989 that it possessed presentness.* This conforms to the theory adumbrated in section 4.5 that uses of "now" directly, rather than indirectly, refer to the B-times at which they are located.

Item (41) is an indicator–ascriber sentence rather than a quasi indicator sentence, since my knowledge of the proposition expressed by (41) does not provide me with a knowledge of the proposition expressed by Alice's noon utterance of "David is now sleeping." It cannot provide me with this knowledge, since (41) is now true and the proposition expressed by Alice's noon utterance is now false; (41) instead provides me with the knowledge that the proposition expressed by her utterance *was* true, at noon. It also provides me with the knowledge that when noon was present, Alice knew a proposition then expressible by a sentence containing a present tensed A-indexical such as "now."

To summarize, Castañeda's theory of the alleged quasi indicator "then" is inadequate because it presupposes the tenseless theory of time. A correct theory of the quasi indicators and indicator–ascribers that represent uses of A-indexicals entails that these quasi indicators and indicator–ascribers *are not B-expressions (as is "then") but A-expressions.* This requires that the properties of quasi indicators and indicator–ascribers be (re)defined in the ways mentioned.

4.7 The Cognitive Significance of A-Indexicals

The problem of the cognitive significance of A-indexicals was not solved by the tenseless "direct reference" theories of A-indexicals, as I argued in chapter 2 regarding Kaplan's, Nathan Salmon's, and Castañeda's theories. The failure to solve the problem of cognitive significance is not unique to these theories or even to extant "direct reference" theories of indexicals in general but is much more widespread. Indeed, it may be fairly said (and is generally acknowledged) that *the problem of cognitive significance* is the major problem confronting the "direct reference" theory of linguistic expressions (of all sorts) and that it has so far resisted solution. My remarks in this section may be understood as an attempt to formulate a partial solution to this problem, that is, a solution to this problem as it pertains to A-indexicals.

It is encumbent upon me to attempt a solution to this problem, since the theory of A-indexicals I developed in section 4.5 *is* a "direct reference" theory (albeit not a "direct reference" theory of the standard sort, which is a tenseless theory). The tensed theory I advocated implies that tokens of "today," "now," and other A-indexicals *refer directly* to moments of time, such that they contribute to the proposition expressed not a sense describing this moment but this moment itself. The propositions expressed by sentence-tokens containing A-indexicals are *de re* propositions, not *de dicto* propositions, since they include as constituents certain particulars, the moments M or M_0, and so on. Of course, I hold that indexicals also contribute a property to this proposition—*presentness* in the case of "today" and "now"—but that is consistent with holding that these indexicals refer directly. My theory is not that these indexicals refer to a moment via the sense *presentness* but that they refer directly to the moment and ascribe to this directly designated moment the property of presentness. A simple property by itself, such as presentness, refers to nothing. What refers are propositional definite descriptions, such as *whatever moment has presentness;* and my theory holds that the property of presentness—but not a propositional definite description—is expressed by the indexical. My position requires me to hold that the indexical's reference to the moment is secured by some means other than its expressed sense—specifically, by the rule that the indexical refers directly to the moment at which it is uttered (in the case of "now" or "today"). Thus, my theory, while a "direct reference" theory, is not a purist version of this theory, since it holds that singular terms, at least A-indexicals, express nonreferring senses *in addition* to referring directly to items.

Now it might be thought, at first glance, that the sense I allow directly referential indexicals to express functions as the cognitive significance of these indexicals. But this sense at best functions as a part of their cognitive significance. A use of "now" at moment M expresses the semantic content *M, which has presentness;* and the issue of the cognitive significance of this use pertains to how we grasp the particular time M, the "manner of presentation" of M. The property of presentness is not a manner of presentation, since such manners are *propositional definite descriptions*. It is widely and correctly assumed in the literature that cognitive significances are *propositional definite descriptions,* although (as I shall argue) this assumption is correct only if these descriptions are understood in an unusually wide sense as including *de re* descriptions, as well as *de dicto* ones. (*De dicto* propositional definite descriptions include only senses as constituents, *de re* descriptions, particulars, as well.)

We have already considered two *de dicto* propositional definite descriptions of the moment M that might seem to some to be candidates for the cognitive significance of "now" or "today," that is, *whatever moment is present* and *whatever moment is actually present*. But the assumption that one of these two descriptions is the cognitive significance of "now" or "today" has the unwelcome consequence that in the case of utterances with the relevant modal and temporal operators, the sentence-token (and proposition) expressed is true, but the associated belief of the speaker is false. If I utter on Thursday, moment M, "It will be true tomorrow that today is Thursday," my utterance is true, since its operand expresses the semantic content *M, which is present, is Thursday.* But my belief is false, since my belief is that *it will be true tomorrow that whatever moment (actually) has presentness is Thursday,* and it will not be true tomorrow that the moment that has presentness (tomorrow) is Thursday.

The line of argument just presented has deeper ramifications than those concerning the "direct reference" theory of A-indexicals, since the form of this argument may be used against most of the extant theories of the cognitive significance of directly referential expressions. Proponents of these theories show that the relevant expressions (indexicals, proper names, natural kind terms, etc.) possess various semantic properties that they would not possess if they expressed the *de dicto* propositional definite descriptions that "indirect reference" theorists allege they express. For example, if "Aristotle" expressed *whoever is the first philosopher to distinguish clearly the four causes,* then "Aristotle did not distinguish the four causes" would be self-contradictory, whereas, in fact, it is merely contingently false. However, it is precisely such *de dicto* propositional definite descriptions that many "direct reference" theorists identify as the cognitive significance of singular and natural kind terms, which has the effect of reintroducing the problem at the level of the speaker's beliefs. If the cognitive significance of "Aristotle" is or includes *whoever is the first philosopher to distinguish clearly the four causes,* then the proposition believed by the person who utters the sentence in question is self-contradictory, which produces the unpalatable consequence that while the proposition expressed by his utterance is contingently false, the proposition that constitutes the cognitive significance of his utterance is necessarily false. This is problematic, because it would not be an isolated case but would introduce a systematic nonparallelism between the speaker's beliefs and the propositions expressed by his utterances. Such a systematic nonparallelism is unacceptable because it goes against "our linguistic intuitions" just as much as do the unpalatable semantic consequences that the "direct reference" theorists showed to be entailed by the "indirect reference" theories.

The difficult task confronting "direct reference" theorists is *to find propositional definite descriptions that function as cognitive significances but that are not the ones they argued could not be expressed by the relevant locutions.* But this is only half the difficulty. The other half is to find a reason for maintaining that these new and unproblematic cognitive significances are not parts of the propositions expressed by the sentence-utterances; for if these significances avoid all the problems in question, then there would seem to be no reason for denying that they are parts of the propositions expressed and therefore that the "indirect reference" theory is true.

I believe this problem is soluble (at least for A-indexicals) only in one way, namely, by taking *experiences* of the speaker as the relevant cognitive significance. Suppose the cognitive significance of my use of "now" is *whatever moment has presentness and actually includes E (among other events),* where "E" refers directly to the complex psychological event composed of my present experiences. This significance is not a *de dicto* propositional definite description, since it is not composed of universals only but also includes particulars, namely, the particular psychical events I am now undergoing, such as my sensing redly, sensing painfully, and imagining bluely, and my acts of thinking. Usually, objects of attitudes *de re* are held to be grasped only by way of *de dicto* propositional definite descriptions; but the objects of the *de re* attitude I have in mind are immediately graspable, since the particulars they involve, by virtue of being *my experiences,* are immediately graspable by me. This enables us to meet a desideratum that must be met by any candidate for cognitive significance, namely, that the cognitive significance *be an object of a knowledge by acquaintance,* rather than *an object of a knowledge by description* (to borrow Russell's phrases but not necessarily all the details of his theory). Objects of acquaintance are either universals or the particulars that are my immediate experiences; thus, cognitive significances must be one or the other or a combination of both.

Before spelling out the details of this conception of *de re* cognitive significance, let us pause to further situate this theory vis-à-vis the classical "indirect reference" theory and the standard "direct reference" theory. The theory of A-indexicals that I am advocating differs from these two theories on *both* semantic and cognitive grounds. I reject the classical indirect reference theory because it says that the semantic content of uses of "now" are *de dicto* contents (senses only, whether this sense be a B-sense or, with Schlesinger and Tichy, an A-sense); and I reject the standard direct reference theory of "now," since it holds that the semantic content of "now" includes no sense but only a particular B-time. My "mixed theory" holds that uses of "now" have for their semantic content a particular time and a sense. Likewise, my "mixed theory" of the cognitive content (significance) of uses of "now" rejects the standard theory that this content is a *de dicto* content. I hold it is a *de re* content, one that includes particulars—although it also includes senses, such as *whatever moment has presentness.*

But I have indicated so far that the *de re* theory of cognitive significance that I advocate meets only one of the desiderata of a theory of cognitive significance, namely, that the cognitive content be an object of acquaintance. It also needs to be shown that this theory of the *de re* cognitive content of uses of "now" enables beliefs to parallel the sentence-utterances in their truth values and the modality of their truth values (e.g., as contingently or necessarily true or false). Consider sentences (19) and (23) from section 4.5:

(19) World War III did not begin today, but it might have.

(23) World War III did not begin at M, which is present; but it might have.

Item (23) expresses the semantic content of an utterance of (19) at the moment M, which (I supposed) has the date property of being May 1, 1986. The problem of cognitive significance is the problem of the user of (19) associating with his use of (19) a cognitive content that has the same truth value as the semantic content. Call the use of (19) at M by John the use U. The use U *semantically* expresses the proposition semantically expressed by (23) but *cognitively* expresses some other proposition that is the cognitive significance of U. Let "E" refer directly to the complex psychological event composed of the experiences of John at M with which John is acquainted. As John utters U, he grasps the *de re* proposition cognitively expressed by U, which is the proposition that we may semantically express by

(1) World War III did not begin at whatever moment is present and actually includes E, but it might have.

Now John does not token (1) at M; he tokens only (19). But his tokening of (19) expresses two propositions, one semantically and one cognitively; and the one it cognitively expresses *can be semantically expressed by (1)*, even though John does not so express it. The *de re* proposition expressed by U and (23) is true if and only if there is some possible world in which M has presentness and includes the start of World War III. But these are precisely the truth conditions of the *de re* proposition expressed by (1); for at M "whatever moment is present and actually includes E" refers *rigidly* to M, by virtue of the modal adverb "actually." The moment that actually includes E is M, and so (1) refers to M in other worlds, as well, even those in which M does not include E or other events that it actually includes. In every world in which it exists, M possesses the property of *actually including E* (i.e., of *including E in the actual world*). If we suppose it is now M, U, (23), and (1) are all true, since M does not include the start of World War III, is present, actually includes E, and in some other possible world is both present and includes the start of World War III.

This account of the cognitive significance of A-indexicals also explains their significance in temporal contexts. Consider an utterance T on Thursday of the sentence

(16) It was true yesterday that today is Thursday.

My account of (16) in section 4.5 implied that this token of "today" refers directly to the day of its utterance, Thursday, and not to the day referred to by the temporal operator, Wednesday. This insensitivity to temporal operators is also preserved by the cognitive content expressed by the token of "today is Thursday"; for this content is semantically expressible by "whatever day is present and actually includes E," and the day that includes E (the psychological experiences of John at the time he utters T) is the day of utterance, Thursday.

It is obvious that the cognitive contents of uses of temporal indexicals also have truth value matches with their semantic contents in the case of extensional utterances; for "M, which is present, is Thursday" clearly refers to the same day as a simultaneous token of "whatever day is present and actually includes E is Thursday," given that "E" directly refers to the experiences of its utterer.

This analysis shows how this theory of cognitive significance enables the truth values of the speaker's beliefs to parallel the truth values of the speaker's semantic contents and thereby to resolve the first main part of the problem of cognitive significance. The second part is to explain why this cognitive content does not do double duty as the semantic content of the utterance. What reason is there to suppose that *M, which is present* is the semantic content of the relevant uses of "now" or "today," rather than *whatever moment is present and actually includes E?* I think the answer is provided by the inclusion of E in the cognitive content. I would note, to begin with, that the answer *cannot* be a reason of the same form as the "direct reference" theorists' reasons for rejecting the idea that the traditional *de dicto* contents are semantic contents. These reasons have the form, *C cannot be the semantic content of locution L, since that would entail that the sentences to which L belongs have (in certain contexts of utterance) truth values or modal properties of truth values that they do not, in fact, have.* The reasons cannot have this form, since the supposition that they do entails the problem of the systematic nonparallelism between the truth-value properties of the speaker's beliefs and utterances. The reason must, instead, be based on the peculiar nature of the *de re* cognitive contents, namely, that they are *private to the speaker* and thereby are not publicly and intersubjectively expressible contents of sentence-utterances. If "E" directly refers to my experiences at moment M, the reason that *whatever moment is present and actually includes E* is not the semantic content of my use of "now" at M is that this content would be expressible only privately, by myself. If I express, instead, the semantic content *M, which is present,* I express a content *also* expressed by anybody else's simultaneous utterance of "now." When other people simultaneously utter "now," they express the same semantic content that I express but grasp this content in a different way, via their own private cognitive contents. Our simultaneous utterances of a sentence such as "It is now noon" semantically express one and the same proposition but cognitively express different propositions, each of the latter being graspable only by the person whose utterance cognitively expresses it. In summary, the reason that the cognitive content is not the semantic content is that rules of ordinary usage are rules of a *public language* that guarantee the *intersubjective expressivity of semantic contents.*

It would be an interesting task to apply this theory of cognitive significance to "here" and "I," proper names, natural-kind terms and other directly referential expressions; but this task falls beyond the purview of the present treatise. The aim of part 1 has been to develop a tensed theory of A-indexicals and other A-expressions with the aim of using this theory to build a more comprehensive theory of language and time, namely, presentism. The presentist theory implies that presentness belongs to the semantic content of tenseless sentences, as well as tensed ones, and that presentness inheres in events absolutely, rather than relative to a reference frame. The current chapter contains my tensed theory of A-locutions, and the results of this chapter will be used in part II as some of the premises from presentism shall be derived.

Notes

1. McTaggart argued that B-propositions and A-propositions are logically incoherent and that only C-propositions (corresponding to McTaggart's C-series) are true. Although C-propositions are different from A- and B-propositions, McTaggart did not claim that A-sentences (as ordinarily used) express C-propositions.

2. Cf. D. H. Mellor, "Tense's Tenseless Truth Conditions," *Analysis* 46, no. 4 (1986): 167–72; Graham Priest, "Tense and Truth Conditions," *Analysis* 46, no. 4 (1986): 162–66; idem, "Tense, *Tense,* and TENSE," *Analysis* 47, no. 4 (1987): 184–87.

3. D. H. Mellor, *Real Time* (Cambridge: Cambridge University Press, 1981), 170.

4. Priest, "Tense, *Tense,* and TENSE," 186.

5. Pavel Tichy, "The Transiency of Truth," *Theoria* 46(1980): 164–82; George Schlesinger, *Aspects of Time* (Indianapolis: Hackett, 1980), 134–35.

6. Alvin Plantinga, "The Boethian Compromise," *American Philosophical Quarterly* 15(1978): 129–38.

7. *Philosophical Studies* 52(1987): 77–98.

8. Quentin Smith, *The Felt Meanings of the World: A Metaphysics of Feeling* (West Lafayette, Ind.: Purdue University Press, 1986), 155–60.

9. Quentin Smith, "Kant and the Beginning of the World," *New Scholasticism* 59(1985): 339–46.

10. Sydney Shoemaker, "Time Without Change," *Journal of Philosophy* 66(1969): 363–81; Richard Swinburne, *Space and Time,* 2d ed. (New York: St. Martin's, 1981); W. H. Newton-Smith, *The Structure of Time* (Boston: Routledge & Kegan Paul, 1980).

11. For a more detailed discussion of this issue, see my "New Theory of Reference Entails Absolute Time and Space," *Philosophy of Science* 58(1991): 411–16.

12. Hector-Neri Castañeda, "The Semiotic Profile of Indexical (Experiential) Reference," *Synthese* 49(1981): 281.

13. Ibid., 291.

14. Idem, "Indicators and Quasi-Indicators," *American Philosophical Quarterly* 4(1967): 93.

15. Ibid., 85.

II

THE ARGUMENT FOR PRESENTISM

5

Presentness as a Logical Subject of A-Sentences

5.1 The Theory of Presentism

In the preface I characterized presentism as a theory implying that every possibly true sentence is synonymous with a sentence of a form such as "Presentness inheres in such-and-such." This initial characterization can now be given a more precise formulation. Presentism is the theory *that every possibly true sentence has present-ness for a logical subject and that every state of affairs has presentness for a metaphysical subject.* The full meaning of this formulation will be explicated in the course of the arguments in part II, but a partial elucidation can be accomplished in this section.

Presentness is a logical subject of a sentence if and only if it is a referent of a part of that sentence (as used on some occasion) and is ascribed a property by a part of that sentence. It seems at first glance that it is obviously false that presentness is a logical subject of every sentence or even of every A-sentence. If the tensed theory of time is true, then presentness is ascribed by the A-sentence "John is walking." But surely, the fact that presentness is ascribed by this sentence does not entail that presentness is a referent of (a part of) this sentence and is itself ascribed some property by this sentence. In this chapter, I shall argue that this entailment does indeed obtain. The properties ascribed to presentness are second-order properties of the form ( ) *inheres in such-and-such* or ( ) *inheres always in such-and-such*. These are not "ordinary" properties, such as roundness or redness; but they are properties, nonetheless (or so I shall argue). It is because, and only because, presentness is a logical subject of every sentence that every sentence is synonymous with a presen-tist sentence, that is, a sentence of a form such as "Presentness inheres in such-and-such." I call these sentences presentist sentences since they•make syntactically explicit the fact that presentness is a referent of a part of the sentence and is ascribed a second-order property by a part of the sentence. The abstract noun "Presentness" is the grammatical subject of these sentences, which makes syntactically explicit the reference to presentness; and a verb phrase of a form such as "inheres in such-and-such" is the grammatical predicate, which makes syntactically explicit the attribution to presentness of a second-order property.

Presentism implies not merely that presentness is a logical subject of every sentence but also that it is a metaphysical subject of every state of affairs. A *state of affairs,* or fact, is a truth-maker of propositions (and derivatively, of sentence-tokens that express propositions); a state of affairs is what makes propositions (and sentence-tokens) true. For every true proposition, there is a state of affairs that corresponds to that proposition and makes it true; a false proposition is one to which there corresponds no state of affairs. The same holds for sentence-tokens. If a sentence-token ascribes a property F to some existent E and the token is true, then there corresponds to the sentence-token a state of affairs that consists of the existent E and the property F as ordered by exemplification. The state of affairs that is E's-exemplification-of-F corresponds to the true token of "E is F" and to the expressed proposition *that E is F.* A *metaphysical subject* of a state of affairs is a "subject of predicates" in a metaphysical sense, where "predicates" means properties and "subject" means possessor of properties. By a metaphysical subject I mean not a particular or substance but any existent that belongs to a state of affairs and has the role or place in that state of affairs of being a property possessor. Substances (persons, chairs, stars) are metaphysical subjects; but they are not the only metaphysical subjects. *Properties* can belong to states of affairs and can have the role or place in them of being property possessors. I shall argue that presentness is a metaphysical subject of states of affairs—indeed, a metaphysical subject of every state of affairs—and has the role or place in states of affairs of possessing a property of the form *inheres in such-and-such.* Consider the state of affairs that corresponds to a true token of "The storm is present." This state of affairs consists of the storm and presentness as ordered to each other by exemplification: the-storm-exemplifies-presentness. This entails that *the storm* is a metaphysical subject of this state of affairs. But states of affairs typically have more than one metaphysical subject, and presentness is also a metaphysical subject of this state of affairs. If the storm exemplifies presentness, that entails that presentness has the second-order property of *inhering in the storm* (i.e., of *being exemplified by the storm*). The state of affairs the-storm-exemplifies-presentness is the same state of affairs as presentness-inheres-in-the-storm. Explanations and arguments for theses such as these will be taken up at length.

But it may be noted here that not every metaphysical subject is also a logical subject, since "reality overflows the use of language." In other words, there are some states of affairs to which no sentence-tokens correspond. Last night there obtained the state of affairs of the wind rattling the shutters, but nobody uttered the sentence "the wind is rattling the shutters." If there is, in some sense, a "metaphysical–semantic isomorphism," the isomorphism is not between states of affairs and sentence-tokens but between states of affairs and propositions. Every state of affairs corresponds to some true proposition and every true proposition corresponds to some state of affairs (as will be argued). This isomorphism between states of affairs and true propositions implies that not every true proposition is expressed by a sentence-token, which is a claim for which I argued in chapter 3.

Presentism implies that presentness is *the universal metaphysical subject,* that is, the subject of every state of affairs. This formulation suggests a further implication of presentism than the aforementioned implication that "presentness is a logical subject of every sentence and a metaphysical subject of every state of affairs." Presentism entails that presentness *and only presentness* is a logical subject of every sentence and a metaphysical subject of every state of affairs. It is not denied that sentences have other logical subjects and states of affairs other metaphysical subjects, but it is maintained that *only presentness* is a subject of *every* sentence and state of affairs. We may express this by saying that only presentness is a *universal* logical subject and a *universal* metaphysical subject and that other existents are merely *local* logical and metaphysical subjects. The storm is a logical subject of "The storm is present," and it is a metaphysical subject of the state of affairs the-storm-exemplifying-presentness; but it is not a logical subject of "John is walking" or of the corresponding state of affairs and therefore is not a universal logical and metaphysical subject.

The idea that presentness is the only universal logico-metaphysical subject is not an intuitively plausible thesis. If it were an intuitively plausible thesis, there would be little need to devote a whole book to arguing for it. I would even admit that presentism is an initially implausible theory, if only for the reason that it seems, at first glance, that any argument that would show that presentness is a *universal* logico-metaphysical subject would also show that many other properties, including all the trivially essential properties (being self-identical, being something, etc.) are also universal logico-metaphysical subjects. If presentness is a logical subject of "Two plus two equals four" (which initially seems dubious), then it would seem that there should be a parallel argument showing that self-identity, being an object, and so on are also logical subjects of this sentence. However, I shall show that initial appearances are deceptive and that there are sound arguments that presentness and only presentness is a logical subject of every sentence and a metaphysical subject of every state of affairs. Part II shall build upon the conclusions of part I and establish that reality is semantically and metaphysically unified in presentness, that is, that presentness alone among existents is such that every sentence-token is about it and alone among existents is such that every state of affairs is a state of it.

The final significant implication of presentism for which I shall argue is that presentness inheres in events absolutely. This is implicit in the remarks just made; for I characterized states of affairs as having such forms as *presentness-inheres-in-such-and-such,* rather than *presentness-inheres-in-such-and-such-relatively-to-a-reference-frame-R.* Presentism entails there is a single "absolute tide of becoming" that encompasses everything, rather than a multiplicity of crisscrossing tides of becoming, one tide for each reference frame. This requires that it be shown that Einstein's Special Theory of Relativity is not about time but something else, namely, the observable luminal relations among physical events.

The thesis that every state of affairs has the form *presentness-inheres-in-such-and-such* gives a precise sense to the familiar claim (e.g., of Kant, Husserl, and Heidegger) that time is the fundamental structure of reality or that everything is unified in temporal presence. It is by virtue of presentness inhering in something that states of affairs are constituted and that there are states of affairs rather than nothing at all. But this is not to say that there might have been no states of affairs.

Presentness inheres contingently in some beings, such as the event of a tree's falling, but inheres necessarily in other beings, such as the equality of four to two plus two. In every possible world, presentness inheres in the equality of 4 to $2 + 2$ (as I will argue in chapter 6) and thus there is the state of affairs constituted by this mathematical presence in every possible world. The point is, rather, that there are states of affairs only by virtue of there being presences (presentness's inherence in something) and that every state of affairs consists of presentness and a property of presentness. Reality is an infinite presence, that is, it consists of presentness inhering in an infinite number of beings. If reality is a wheel, then presentness is its hub and its spokes are properties of presentness, properties of the form *inheres-in-such-and-such*. Since presentness is the only universal metaphysical subject, presentness is metaphysically preeminent among all concrete and abstract objects, and thus there is a sense in which presentness plays a role in presentist theories that is analogous to the role that the property of oneness (the One) plays in Neoplatonism and that the property of goodness (the Good) plays in Platonism. I shall show in the conclusion of this book that appreciating *that presentness inheres in the infinite totality of beings* is tantamount to the appreciating the maximal extent and unity of all reality.

My argument for presentism is divided into three chapters. In this chapter, I argue that presentness is a logical subject of every A-sentence. This argument requires the development of a theory of *de re* and *de dicto* propositions, which takes up most of the chapter. I conclude by showing that the thesis that presentness is a logical subject of A-sentences does not entail McTaggart's paradox.

In chapter 6, I argue that all B-sentences, natural law sentences, mathematical sentences, and other tenseless sentences also refer to presentness and ascribe to it some property. I show that the so-called tenseless copula is in fact "tensed" in a broad sense. It ascribes not an A-property (presentness, pastness, or futurity) but such properties as *everlasting presentness* (is present and will always be present) or *sempiternal presentness* (was always present, is present, and will always be present). This chapter includes an argument that everything exists in time, including abstract objects. It also includes an argument that nothing other than presentness is a metaphysical subject of every state of affairs.

In chapter 7, I argue that presentness inheres in events or states absolutely, not relatively to some reference frame. The Special Theory of Relativity, I argue, entails that certain observable luminal relations are relative to reference frames but does not entail that temporal relations or properties are relative to a reference frame.

We begin in the present chapter with an argument that presentness is a logical subject of A-sentences. This requires an initial analysis of the semantic concept of *property ascription*.

5.2 A Preliminary Analysis of Property Ascription

Suppose that uses of "John is running" ascribe presentness to John's running but neither refer to presentness nor ascribe to it the property of *inhering in John's running*. Suppose that "John is running" *entails* but is not *synonymous* with some sentence that did have presentness for a logical subject, such as "Presentness inheres in John's running." It would follow that there are some sentences that are not *about* presentness and therefore that reality is not semantically unified in presentness. It would follow, further, that the state of affairs corresponding to a true token of "John is running" consists of a state of John, namely, John's present possession of running, but not of a state of presentness. The state of affairs consisting of a state of presentness, presentness's possession of *inhering in John's running*—would be required by the state of John, such that the latter could not obtain unless the former obtained; but they would be different states of affairs (according to the identity criteria of states of affairs I will present in section 6.1), and this difference would entail that reality is not metaphysically unified in presentness. Presentism would be false. I believe, however, that there is a line of argument that shows that "S ascribes presentness" entails "S has presentness for a logical subject" and therefore that uses of "John is running" are *synonymous* with uses of "Presentness inheres in John's running."

F. P. Ramsey wrote in 1925 that "it seems to me as clear as anything can be in philosophy that the two sentences 'Socrates is wise', 'Wisdom is a characteristic of Socrates' assert the same fact and express the same proposition."[1] But Ramsey contented himself with this proclamation of obviousness and made no attempt to argue for this claim. It seems to me that a claim of this sort is not obviously true. It may be granted that it is obvious that Ramsey's two sentences are *logically equivalent,* but not that they are *semantically identical,* that is, express one and the same proposition. I take the fact of their semantic identity to be an *unobvious truth* that requires considerable argumentation to establish. But I am somewhat misrepresenting the issue that faces presentism by introducing it in terms of Ramsey's claim. Ramsey's claim is very close to, but not identical with, the claim pertinent to presentism. Ramsey is saying that a sentence with an adjective ("wise") is synonymous with a sentence whose grammatical subject is a nominalization of the adjective ("wisdom"), whereas the presentist claim is that a sentence with a copula that is in the present tense is synonymous with a sentence that has for its grammatical subject a nominalization (so to speak) of that tense. "Presentness" may be said to be a nominalization of the present tense of the "is" in "John is running" in the sense that it is a name of the property to which the present tense of the "is" semantically relates, (or so I shall argue). Let us begin with some preliminary distinctions.

In terms of the terminology I used in part I, my presentist thesis is that a token of the ordinary sentence "John is running" is *translated* by a token of the presentist sentence "Presentness inheres in John's running," where "translates" means (i) has the same truth conditions as, (ii) has the same confirmation conditions as, (iii) is logically equivalent to, and (iv) is logically identical with. Since the first three conditions are necessary but insufficient conditions of translation and the fourth condition, logical identity, is both necessary and sufficient, I shall concentrate on logical identity. Two sentence-tokens are logically identical if and only if they refer in the same way to the same items and ascribe to these items the same n-adic properties. Two sentence-tokens refer in the same way to something if and only if they both either (a) refer to it directly or (b) refer to it via the same propositional definite or indefinite description. A sentence-token refers to something if and only if some part of that sentence-token—say, the token of a proper name—refers to something. The relation that a sentence-token as a whole bears to something is what I call *correspondence;* if true, a sentence-token corresponds to a state of affairs.

We may now formulate our question more precisely: Do tokens of "John is running" and "Presentness inheres in John's running" both *refer* to presentness and refer to presentness *in the same way*? Do they both *ascribe* to presentness *the same property*? Are they logically identical in all other respects, as well?

Some philosophers might contend that uses of "Presentness inheres in John's running" refer to presentness, since they contain an abstract name of presentness but that uses of "John is running" do not refer to presentness, since the latter sentence contains no name or description of presentness. Manifestly, the present tense of the copula in "John is running" is neither a name nor a description. The proof that only names and descriptions *refer* (the philosophers may continue) is that the denial of this thesis entails "the paradox of the list." John Searle states a version of this "proof" in his argument against the claim that predicates refer, but an analogous version can be directed against the presentist claim that tenses refer. Searle's version goes as follows:

> If one is given two premises, which all the philosophers [who hold that predicates refer] hold implicitly or explicitly, one can derive a *reductio ad absurdum* of the thesis that it is the function of predicates to refer. The premises are
>   1. The paradigmatic cases of reference are the uses of singular referring expressions to refer to their referents.
>   2. Leibniz's law: if two expressions refer to the same object they are inter-substitutable *salva veritate*.
> Combine these with the thesis:
>   3. It is the function of predicate expressions, like singular referring expressions, to refer.
> Then in any subject-predicate proposition of the form "fa" arbitrarily assign any proper name "b" to the referent of the predicate, and by substitution we can reduce the original sentence to a list: "b a," which is not even a sentence.
> At this point there are two possible maneuvers. One can say:
>   (a) The sense of "refer" (and hence of "stand for", "designates", and all the rest) is different for predicates from what it is for singular referring expressions. Hence the reduction to a list is invalid.
>   (b) The entity referred to by the predicate is a very peculiar entity, so peculiar that as soon as we try to refer to it with a referring expression (*Eigenname*) we find ourselves referring to an entity of a different kind. Hence it is impossible to assign a name to it, and the reduction to a list is invalid.
> Frege in effect adopted (b). Neither of these attempts to avoid the breakdown is satisfactory. Maneuver (a) leaves the notion of referring in the case of predicates wholly unexplicated and amounts in effect to a surrender of the thesis at issue, since in the statement of the thesis the relation of a singular referring expression to its referent was presented as the paradigm of referring. Maneuver (b) again is surrounded by mystery and incomprehensibility, and apart from that it produces a formal contradiction as soon as we apply a general term to the kind of thing which is referred to by the predicate—a contradiction of the form, e.g., the concept *horse* is not a concept.[2]

Searle concludes from this that predicates do not have the function of referring. An analogous version of this argument can be constructed for the present tense. If the present tense refers to presentness, then a sentence such as "John is running" can be replaced by a sentence with no present tense but a name for the property to which the present tense refers, so that "John is running" becomes "John's running present-ness," which is not even a sentence.

Leibniz's law only applies to extensional contexts, but I will waive this objec-tion in responding to Searle. The pertinent weakness in his argument is that he tacitly assumes that "uses of expression E refer" entails "uses of expression E refer and have no other semantic function." This assumption shows up in the way he formulates thesis 3, that it is *"the* [rather than *a*] function of predicate expressions . . . to refer" as if their possession of this function required that they possess no other function. It is true of uses of singular referring expressions that they refer and have no other semantic function, but this is not true of grammatical predicates and tenses. A third alternative to Searle's (a) and (b), an alternative (c), is that grammati-cal predicates and tenses refer to properties *and have an additional semantic func-tion that enables a sentence, rather than a list, to be formed.* This additional semantic function I shall call *conveying that* the property to which they refer *stands in a propositional relation* to the semantic contents of the rest of the sentence-token. This alternative, which I shall develop and defend in this and the following sections, is (as applied to the present tense) that "the present tense of the copula *ascribes* presentness to John's running" is to be analyzed as meaning "the present tense of the copula *refers (directly) to presentness and conveys that it is propositionally related to John's running.*" The unfamiliar terms here are "conveys that" and "is proposi-tionally related to," and they shall be defined in what follows. My goal is to show that the ordinary sentence "John is running" is synonymous with the presentist sentence "Presentness inheres in John's running" by virtue of the facts that (i) the present tense's *direct reference* to presentness is captured by the grammatical subject "Presentness" of the presentist sentence and (ii) the present tense's *conveyance* of the propositional relatedness of presentness to John's running is captured by the expression "inheres in" in this presentist sentence.

The most important element in the definition of the *propositional relation* in-volves a reference to truth values. If several items stand in the propositional relation to each other, then these items as standing in this relation have a truth value. Sentence-tokens have truth-values derivatively, by virtue of semantically expressing items as standing in the propositional relation. Thus, this first element in the definition of a propositional relation is more exactly stated by saying that if several items are related by this relation, then these items as so related *originarily (non-derivatively)* possess a truth value. The difference between the heap of individuals and properties,

(a) John, the property of running, the property of presentness

and the *de re* proposition expressed by a token of "John is running" is that the *de re* proposition consists of the three items mentioned in (a) *as related by the proposi-*

tional relation. It is the function of a sentence-token both to pick out the items that are propositionally related and to indicate that they are propositionally related.

The specific nature of the propositional relation as an original truth-vehicle-maker can be further demarcated by showing the respect in which the widely adopted practice of taking propositions to be—or to be represented by—n-tuples is inadequate. I provisionally adopted this practice in chapter 4, but it must be abandoned in the final analysis. Suppose we "represent" the proposition expressed by a use of "It is noon" by the ordered pair.

⟨Noon, the property of presentness⟩.

Now an ordered pair has two defining properties: (a) it is an entity composed of two entities and (b) the two entities are identifiable as *the first* and *the second*. But this relatedness of noon and presentness as the first and second is hardly sufficient to comprise a proposition. Noon's being the first and presentness' being the second is obviously not something that bears a truth value. The use of the n-tuple notation to represent propositions is at best a temporary measure, one that postpones the attempt to determine what propositions really are. Joseph Almog seems to be one of the few contemporary users of this notation who both recognizes and admits this: "I have used the n-tuple notation for propositions. I don't actually think propositions are sequences, in the set theoretic sense. But, lacking a full understanding of the matter at present, I use the n-tuple notation as a temporary measure."

The necessity of introducing the propositional relation can also be seen from a brief examination of Russell's 1903 theory of *de re* propositions. Writing of a singular proposition involving the things A and B, Russell says:

> Consider, for example, the proposition "A differs from B." [The quotation marks designate a proposition, not a sentence.] The constituents of this proposition, if we analyze it, appear to be only A, difference, B. Yet these constituents, thus placed side by side, do not reconstitute the proposition. The difference which occurs in the proposition actually relates A and B, whereas the difference after analysis is a notion which has no connection with A and B. It may be said that we ought, in the analysis, to mention the relations which difference has to A and B, relations which are expressed by *is* and *from* when we say "A is different from B." These relations consist in the fact that A is referent and B relatum with respect to difference. But "A, referent, difference, relatum, B" is still merely a list of terms, not a proposition. A proposition, in fact, is essentially a unity, and when analysis has destroyed the unity, no enumeration of constituents will restore the proposition. The verb, when used as a verb, embodies the unity of the proposition, and is thus distinguishable from the verb considered as a term, though I do not know how to give a clear account of the precise nature of the distinction.[4]

This passage is unacceptable. First of all, it is simply false that "the difference which occurs in the proposition actually relates A and B." Suppose the proposition is false. Then, *difference* does not relate A and B; for A does not then stand to B in the relation of difference. Yet there still is a proposition (and not a mere list of terms) in which *difference* occurs. But even if the proposition is true, it is false that the difference relates A and B *in the proposition*. Rather, *difference* relates A and B in the state of affairs that corresponds to the proposition. In the proposition, instead of A and B standing in the relation of *difference,* A, B, and *difference* stand in some other relation R. But R cannot be the further relation Russell introduces, namely, that expressed by "is" and "from" when we say "A is different from B"; for these relations (if there are such) belong, instead, to the state of affairs that obtains if the proposition is true. The relation R is *the propositional relation;* A, B, and difference, by virtue of standing in the propositional relation, form a truth-valued complex, that is, a proposition.

Russell is incorrect when he writes that "the verb, when used as a verb, embodies the unity of the proposition." Russell is here using "verb" in an eccentric sense, to refer, not to a linguistic entity, but to a relation; and the relations he has in mind are relations such as *differs.* According to Russell, the relation *differs* embodies the unity of the proposition *that A differs from B.* But according to the view I am developing, *differs* is, instead, one of the parts of the proposition that is unified or ordered, and is not itself the unifying or ordering item. The ordering or unifying item is the propositional relation.

This is an occasion for distinguishing between the senses of "parts" and "constituents" (as I use these words). For any given proposition p, there are parts and constituents that belong to p, such that all parts are also constituents but not all constituents are also parts. Something is a part of p if and only if it is a term of a propositional relation that belongs to p. Something is a constituent of p if and only if it is a part of p or a propositional relation that relates the parts of p. In the singular proposition Russell discusses, A, B, and difference are parts and the propositional relation is merely a constituent.

So far, I have defined the propositional relation as (a) the original truth-vehicle-maker and (b) the unifying or ordering constituent of a proposition. Further clarification is achieved if we discuss this relation in the context of a discussion of the semantic property that tenses and copulae possess of *conveying that* this propositional relation obtains.

5.3 The Conveyance Relation

A third element in the definition of the propositional relation is that it is the relation R the obtaining of which is *conveyed* by the tense or copulae of a sentence token (in the cases where the proposition is linguistically expressed). In other words, the obtaining of the propositional relation is a relational term of the semantic relation of *conveyance* whose other relational term is the copula, the copula's tense, or (in cases where the sentence-token has no copula) the verb that incorporates within itself the semantic function of the copula. Since "conveyance" is an unfamiliar term and is itself defined in terms of the propositional relation (as the semantic relation one of whose terms is the obtaining of the propositional relation), elucidation is best achieved by giving an example.

Consider any token T of "John is running." Let us suppose that "John" refers directly to John, that "running" refers directly to the property of running, and that the present tense of the copula refers directly to presentness. What conveys that these items are propositionally related? Two syntactical parts of this sentence-token do this: *the copula* that is in the present tense conveys that the property of running is propositionally related to John; and the *present tense* of this copula, in addition to referring to presentness, conveys that presentness and John's running are propositionally related.

This is made more precise in terms of the following distinction between the copula and its tense. I shall call the *pure copula* that which is common to "is," "was," and "will be" in the sentences "John is running," "John was running," and "John will be running." The pure copula is not the property of *being tensed* but exemplifies this property and, by virtue of exemplifying this property, becomes a *tensed copula*. The *tense* of a copula is just its property of *being tensed,* which is, of course, always a specific property, such as *being present tensed* (or *being in the present tense*). The pure copula in a token of "John is running" conveys that John and running are propositionally related and the present tense of this copula conveys that presentness and John's running are propositionally related. Analogously, the pure copula in "John was running" indicates that John and running are propositionally related and the past tense of this copula indicates that present pastness and John's running are propositionally related.

Further elucidation of the conveyance relation is achieved through pointing out how it differs from the semantic relation of direct reference. If the "is" in a token of "John is running" directly referred to the propositional relation, this token would not express a proposition but a mere heap or aggregate, namely, the heap

(a) John, the property of running, the property of presentness, the propositional relation.

Direct reference is a relation that merely picks out something and serves to introduce it into the proposition. It is not a relation that reveals the order or unity of the proposition. The semantic relation of *conveyance,* on the other hand, depicts that the items introduced into the proposition by the other parts of the sentence-token *stand in a certain relation, namely, the propositional relation.* The pure copula in a token of "John is running" conveys that John and the property of running are propositionally related, as represented in Figure 5.1. The unbroken arrow that connects the token "John" to John and that connects "running" to the property of running stands for the semantic relation of *direct reference*. The broken arrow that connects the token of the pure copula "is" to the propositional relation stands for the semantic relation of *conveyance*. Now the theory of relations uses the terms *referent* and *relatum* to refer to terms of relations, which is a different sense of "referent"

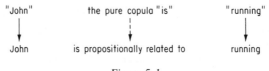

Figure 5.1

than the semantic sense, as when it is said, "The person John is a direct referent of the token "'John'." In the theory of relations, an asymmetrical relation is defined as having a direction; the relation proceeds from the referent to the relatum. For example, the relation *being the father of* proceeds from the father to the child, such that the father is the referent and the child the relatum of this relation.

The relatum of the conveyance relation is formally represented in Figure 5.2. The semantic content (C) of the pure copula "is," the referent of the copula's conveyance relation, remains the same when "John" and "running" are exchanged for other locutions, such as "Alice" and "walking." It also remains the same when the tense is changed, so that C is also the semantic content of the pure copula in "John was running."

I should add here that the copula and its tense are not the only referents of the conveyance relation. In sentences without copula, the verb bears this relation to the obtaining of the propositional relation. In "John loves Jane," the verb "loves" not only directly refers to the relation of *loving* but also has a "copulative function" of conveying that John is propositionally related to the relational property *loving Jane.* "Loves" is here analogous to the tense of a copula, in that it serves both to refer directly to an n-adic property (*loves*) and to convey that this property (or a more complex property of which *loves* is a part, i.e., *loving Jane*) stands in a propositional relation to another item or items.

The semantic relations of the present tense of the copula in a token of "John is running" are represented in Figure 5.3. The unbroken arrow indicates that the present tense directly refers to presentness and the broken arrow indicates that this tense conveys that presentness stands in a propositional relation to John's running. John's running is placed in brackets to indicate that it is a semantic content not of the present tense but, rather, of the other syntactical parts of the sentence-token.

My suggestion is that the present tense of the "is" in the sentence "John is running" is logically identical with the phrase "presentness inheres in" in the presentist sentence "Presentness inheres in John's running." The difference is that in the ordinary sentence a single syntactical element, the present tense of the copula, stands in two semantic relations to two semantic contents (it refers to presentness and conveys its propositional relatedness), whereas in the presentist sentence separate words are introduced to bear separately these two semantic relations. "Presentness" bears the relation of direct reference to presentness, and "inheres in" conveys that presentness is propositionally related to John's running (see Figure 5.4). My suggestion, in short, is that the presentist sentence, as tokened in a context C, is logically identical to the ordinary sentence as tokened in C. They are semantically identical but syntactically different. The difference is that the presentist sentence-

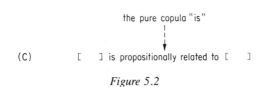

Figure 5.2

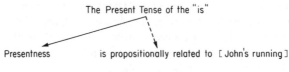

Figure 5.3

token makes *syntactically explicit* the presentist semantic content of the ordinary sentence-token. The presentist semantic content is *presentness-as-propositionally-related-to ( )* and the presentist sentence-token makes this content syntactically explicit in that it has separate words or phrases that bear the relations of *direct reference* and *conveyance* to the respective aspects of this semantic content. There is an additional *psychological* difference in that the presentist sentence-token serves to direct attention upon presentness (by virtue of having "Presentness" for its grammatical subject), whereas the ordinary sentence-token directs attention upon John (by virtue of having "John" for its grammatical subject).

As I have said, I am *suggesting* that this logical identity between the ordinary and presentist sentence-token obtains. I have not yet demonstrated this. I have contributed something toward this demonstration, but further distinctions need to be made and some objections considered. The next task is to explain two different species of propositional relatedness that can be conveyed by tenses and copulae.

5.4 Two Species of the Propositional Relation

The next major step in defining the propositional relation and the associated relation of conveyance is to show that there are two different species of the propositional relation. The fact that there are different species is clearly illustrated by the difference between *de re* and *de dicto* propositions. Suppose that John is the tallest man and that the sentence "The tallest man is running" is used to express a *de dicto* proposition about John. The definite description "the tallest man" introduces a *sense* into the proposition, rather than John himself, so that the expressed proposition includes the following items:

(1) the property of being the tallest man, the property of running, the property of presentness.

These items form a proposition by virtue of standing in a propositional relation, such that tokens of "the tallest man," the pure copula "is," and the verb "running"

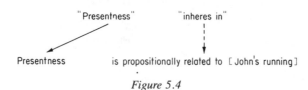

Figure 5.4

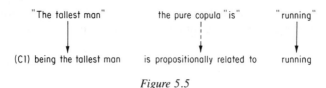

Figure 5.5

are referents of semantic relations as depicted in Figure 5.5. Compare C1 with the semantic relata of the relevant syntactical parts of the token of "John is running,"

(C2) John   is propositionally related to   running.

Now the propositional relation in (C1) cannot be the same propositional relation as belongs to (C2), for the propositional relation in (C2) relates its referent and relatum in such a way that the referent *exemplifies* the relatum if the proposition is true, whereas if the proposition expressed by a token of the "The tallest man is running" is true, the referent of the propositional relation does not *exemplify* the relatum. If the proposition expressed by a token of "John is running" is true, then John *exemplifies* running. But if the proposition expressed by a token of "The tallest human is running" is true, the property of *being the tallest man* does not exemplify the property of running; for properties cannot run. If this proposition is true, it is, instead, the case that the referent of the propositional relation *is exemplified by the same existent that exemplifies the relatum of the propositional relation*. In other words, the property of *being the tallest human* is coexemplified with the property of running. (The existent that possesses the one property also possesses the other property.) Thus, we may define the propositional relation that occurs in (C2) as the relation E, such that

(E) If A is the referent of E, and B the relatum, then A *exemplifies* B if and only if the proposition is true.

The propositional relation C that occurs in (C1) is provisionally defined as follows:

(C) If A is the referent of C, and B the relatum, then A *is coexemplified with* B if and only if the proposition is true.

The E-relation is asymmetrical, but the C-relation is symmetrical. The referent and relatum of the E-relation are not interchangeable. If A has the E-relation to B, then B (except in unusual cases, as when a property is predicated of itself) does not have the E-relation to B. This is obvious from the example of John and running. John is E-related to running, since if the proposition is true, John exemplifies running. But running is not E-related to John, since if the proposition is true, it is false that running exemplifies John.

Let us apply these notions to presentness. Presentness, I suggest, is C-related, rather than E-related, to other parts of the proposition in both the *de dicto* proposition *that the tallest man is running* and in the *de re* proposition *that John is running*.

It might seem surprising that I claim this is the case in the *de re* proposition, for it might seem that in the *de re* proposition, it is the case that

(1) John's running   is propositionally E-related to   presentness,

since if the proposition is true, then John's running *exemplifies* presentness. But this appearance is misleading, since "John's running" as used to designate a relational term of a propositional relation expressed by a token of "John is running" cannot designate the actual occurrence of John's running, the event of John running; for this event is not a part of the proposition. Admittedly, the proposition is *de re*, but this means only that John is a part of this proposition. The event of his running cannot be a part, for then the proposition could not be false. If the proposition is false, John is not running, in which case there is no such event. Thus, "John's running" as used to designate a relational term of the propositional relation must, instead, designate some complex that contains John and the property of running as ordered to each other in a certain way, such that if the proposition is true, then what corresponds to John as so ordered to running is the event of John's exemplifying running. What is this complex? The preceding discussions of the propositional E-relatedness of John to running gives us the answer: the complex is *John as propositionally E-related to running*. Presentness is C-related to this complex, such that if the proposition is true, then presentness is exemplified by the event that corresponds to this complex, namely, the event of *John's exemplifying running*.

This analysis requires us to broaden the provisional definition C of the propositional C-relation, so that *coexemplification* becomes only one species of the connection to constituents of a state of affairs that obtains if the proposition is true. Exemplification is a tie of a property to a property possessor; but since the propositional complex John-as-E-related-to-running is not a property, it cannot, strictly speaking, be said to be exemplified. This complex is not exemplified by the event that belongs to the corresponding state of affairs but stands in another sort of connection with it, what I call *quasi correspondence* and will define in section 5.6.

Let us introduce the term *satisfies* and say that a referent or relatum of a propositional C-relation is satisfied by a constituent of a state of affairs if and only if it is either exemplified by a constituent of it or quasi-corresponds to a constituent of it. Two propositional relational terms T1 and T2 are thus "cosatisfied" if and only if there is some constituent A of the state of affairs such that T1 and T2 are both satisfied by A (i.e., T1 either is exemplified by, or quasi-corresponds to, A; and B either is exemplified by, or quasi-corresponds to, A). Our new definition of the propositional C-relation is then

(C′) If A is the referent of C, and B the relatum, then A is *cosatisfied with* B if and only if the proposition is true.

Of course, this definition will gain full intelligibility only when we define "quasi correspondence" in section 5.6; but I think that even in advance of this definition, the introduction of the E and C propositional relations enable us to provide *formal identity criteria* for propositions. Two propositions p and p′ are identical if and only if two conditions hold:

1. p includes all and only the same parts as p'.
2. The parts of p are propositionally related in the same ways as the parts of p'.

Thus, *that John is running* is different from both *that Alice is running* and *that the tallest man is running*, since the first proposition alone includes John as a part. The usefulness of the E and C relations is in the employment of criterion 2 to distinguish between propositions that contain the same parts, such as the two propositions

(a) *that unextendedness is a property*

(b) *that being a property is unextended.*

Propositions (a) and (b) differ, since their parts are interrelated in different ways. In (a), *being unextended* is E-related to *being a property* (but not vice versa); and in (b), *being a property* is E-related to *being unextended* (but not vice versa). If (a) is true, it corresponds to *unextendedness'* exemplification of *being a property* and if (b) is true, it corresponds to *being a property's* exemplification of *unextendedness.*

It is instructive to compare these identify criteria of propositions with those introduced by Roderick Chisholm in *Person and Object.* He writes:

> What would be a non-trivial criterion of identity for states of affairs [Chisholm uses "states of affairs" to designate what I call propositions]? Since we are characterizing states of affairs as possible intentional objects, as things which are such that they may be accepted, we can introduce the following strict concept of entailment:
>
> > D.IV.2 p entails q = Df p is necessarily such that (a) it if obtains then q obtains and (b) whoever accepts it accepts q.
>
> And now we may affirm this non-trivial criterion of identity: if a state of affairs p is identical with a state of affairs q, then p entails q and q entails p.[5]

It is instructive to see why Chisholm's criterion is not biconditional. If it were biconditional (such that it read *p is identical with q if and only if p entails q and q entails p*), it would be unacceptable. If it were biconditional, it would be inconsistent with the following trivial criterion (or necessary condition) of identity:

3. Something A is identical with something B only if A includes no part not included by B and B includes no part not included by A.

It is plausible to suppose that there is some proposition p that "entails" (in Chisholm's "strict" sense) q but includes some part that q does not. Chisholm's criterion requires p and q to be logically equivalent and to be such that if one is accepted, so is the other. But it is reasonable to believe that there are some propositions that include some different parts but that are nonetheless logically equivalent and require mutual acceptance. It is plausible that the proposition *that the sun has a size* is not identical with the proposition *that the sun has a shape*. Nevertheless, they are logically equivalent and (arguably) one cannot be accepted unless the other is accepted. Thus, if Chisholm's criterion were biconditional, it would not allow us to admit or explain the difference between these two propositions. But the criteria 1 and 2 do. They allow us to say that the proposition *that the sun has a size* includes, as one of its parts, the property of size but does not include the property of shape and that the converse holds for *that the sun has a shape*. Criteria 1 and 2 are finer-grained than the biconditional version of Chisholm's criterion and, unlike the latter, pertain to the constituents of propositions, rather than to their relations to believers or other propositions. But are criteria 1 and 2 nontrivial? That depends on exactly what one means by "nontrivial criterion of identity," which Chisholm does not define; nor shall I attempt to do so. But I will say that it seems on purely intuitive grounds that criteria 1 and 2 are nontrivial if they are explicated as follows:

> 1A. All the individuals, properties, or other items that are relational terms of propositional relations in p are also relational terms of propositional relations in p'.

> 2A. For any propositional relational terms A and B that belong to both p and p', it is the case that (i) if A is E-related to B in p, then A is E-related to B in p'; or (ii) if B is E-related to A in p, then B is E-related to A in p'; or (iii) if A and B are C-related in p, then A and B are C-related in p'.

These two fine-grained criteria of propositional identity will be used in several of my subsequent arguments for the truth of presentism.

5.5 Direct and Indirect Property Ascription

At this point, the analyses have become complicated to the point where we need to reset our bearings. This is especially required because still further complications are needed before we can reach the goal of concluding that presentism is true. We need to establish (at a minimum) that every ordinary sentence has presentness for a logical subject. The task of the current chapter is to show that every ordinary A-sentence has presentness for a logical subject. My route to this conclusion is via the argument that the tenses of copulae both *directly refer* to presentness and *convey that* presentness (or present pastness or present futurity) is propositionally related to the semantic contents of the other elements of the sentence. At this juncture, I have given some reasons to believe that the present tense does directly refer to present-ness and convey that presentness is propositionally related to other semantic contents. This provides justification for the view that ordinary A-sentences, in respect of their tense, *refer* to presentness. But how does it show that these sentences *ascribe a monadic property to presentness*? This must be shown, for presentness is a logical subject of ordinary A-sentences only if it is both a referent of them and is ascribed a monadic property by them.

I propose to answer this question in terms of the tense's semantic function of conveying that presentness is propositionally related to the other semantic contents of the sentence. I shall argue that the tense's conveyance of this propositional relatedness entails that the sentence *ascribes to presentness a monadic property*.

In order to show this, I need to make a distinction between *direct* and *indirect* property ascription. If a property F is ascribed by a sentence-token, then F is either *directly ascribed* or *indirectly ascribed*. F is directly ascribed if a part of the sentence-token *directly refers to* F. Presentness is directly ascribed by present tensed

A-sentence-tokens, since the present tense of their copulae or verbs directly refer to presentness. "Direct ascription" is the species of ascription we have been working with in the past two sections. A more difficult notion is *indirect ascription*. A property F is indirectly ascribed by a sentence-token if a part of that token *indirectly refers to F*. A property F is indirectly referred to by a locution L if L *refers to F via sense*. The most obvious cases of indirect ascription occur in sentences such as

(1) The book jacket has whatever color is the color of the sky.

Sentence (1) indirectly ascribes *blueness* to the book jacket in that it contains an expression that indirectly refers to blueness. The expression "whatever color is the color of the sky" expresses a descriptive sense and refers to blueness via this descriptive sense.

But there is also a less obvious form of indirect ascription, one that is especially pertinent to presentism. Present tensed A-sentences directly ascribe presentness but *indirectly ascribe to presentness a property of the form ( ) inheres in B*, where "B" stands for some being in the present tensed sense, that is, some thing or event in which presentness inheres. Tokens of "John is running" indirectly ascribe to presentness a property of *inhering in John's running*.

The best way to elucidate this idea is in terms of presentist sentences, such as

(2) Presentness inheres in John's running.

Sentence (2) is not a sentence in ordinary language (at least, it is not meant to be) but a sentence in an artificial language, namely, the *presentist language* I am constructing. Accordingly, the matter at hand is not to *find out*, by a study of the actual use of (2), the semantic properties of (2) but to *stipulate* its semantic properties. (I allow that it is possible that such sentences also have an ordinary use and even that their ordinary semantic properties are exactly those that I stipulate them to have in my artificial language, but it is not necessary for my purpose—to construct an artificial presentist language—to establish that this is actually the case.) I stipulate that "Presentness" directly refers to presentness and that the grammatical predicate "inheres in John's running" indirectly refers to a property that includes the actual event of John's running. The grammatical predicate indirectly refers to this property, since only in this case, can the proposition expressed by a use of this sentence exist and be false. If something is directly referred to by a part of a sentence-token, this referent is introduced into the proposition expressed by that sentence-token. (This seems to me obvious and is accepted, as far as I know, by all proponents of the "direct reference" theory that countenance propositions.) Accordingly, I hold that the grammatical predicate "inheres in John's running" *indirectly refers, by means of expressing a sense*, to the property that actually includes the event of John's running. The expressed sense is a constituent of the proposition. The expressed sense is the semantic content of "inheres in John's running" as depicted in Figure 5.6. My stipulation is that "inheres in" is a copula whose sole semantic function is to *convey that* presentness is propositionally C-related to the propositional complex John-as-E-related-to-running. I also stipulate that "John's running" ex-

Figure 5.6

presses this propositional complex itself. But since "John" directly refers to John, this complex includes an individual, John. So is it really a sense? This depends, of course, on how one defines "sense." If "A is a sense" entails "A includes no individuals as constituents," then S is not a sense. It is, instead, an ordered complex of senses and an individual. But if "sense" means "the semantic content of a locution that is the means by which the locution indirectly refers to something," then S is a sense; for by virtue of expressing S, the grammatical predicate "inheres in John's running" indirectly refers to the property that consists of the event of John's running and the property tie of inherence. If the utterance of the sentence "Presentness inheres in John's running" is false, then "inheres in John's running" expresses a sense but fails of reference. There is no property to which this predicate refers, since there is no event of John's running.

It may be doubted whether the indirect referent of the grammatical predicate "inheres in John's running" is a *property*. After all, Russell's paradox shows that "G is a grammatical predicate" does not entail "There is a property correlating to G" (e.g., the predicate "is non-self-exemplifying," correlates to no property). And surely a property that consists of an event and a property tie of inherence is a rather bizarre candidate for a property, hardly of the garden variety like "wisdom." But these points are consistent with a plausible definition of "property" that enables this to count as a property, as I shall show in the next section.

Given this stipulated semantic content of (2), what reason do we have to think that the ordinary A-sentence "John is running" has the same content? I believe the account just given of this A-sentence entails that it has this content. I have argued that the present tense of the copula directly refers to presentness and conveys that it is C-related to the sense of "John's running," where this sense is John-as-E-related-to-running. This is precisely the semantic content of utterances of (2). Thus, an utterance of (2) in a given context of use expresses the same proposition as an utterance of "John is running" in that context. I believe that this goes a long way toward showing that presentism is true at least for ordinary present tensed A-sentences, for it gives us some reasons to believe that these sentences all directly refer to presentness and indirectly ascribe to it a monadic property of the form *inheres in B*.

This applies to *de dicto* present tensed A-sentences, as well, although this may not be obvious at first glance. Consider the presentist sentence, "Presentness inheres in the tallest man's running." Here the grammatical predicate "inheres in the tallest man's running" has the semantic relations depicted in Figure 5.7. To say that *being the tallest man* is propositionally C-related to *running* is to say that if the proposition is true, these two properties are cosatisfied. In this case, cosatisfaction is coexemplification; if the proposition is true, *being the tallest man* is exemplified by

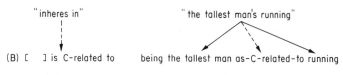

Figure 5.7

whatever exemplifies *running*. Now the actual event of the tallest's man's running is a constituent of the property that is the indirect referent of "inheres in the tallest man's running," the property *inhering in the tallest man's running,* where this property includes the tallest man himself as exemplifying running. This property includes the individual, John (supposing he is the tallest man), as exemplifying both *being the tallest man* and *running.* Since "Presentness inheres in" in "Presentness inheres in the tallest man's running" has the same semantic content as the present tense of the copula in "The tallest man is running," it follows that this ordinary *de dicto* sentence also indirectly ascribes to presentness the property *inhering in the tallest man's running.*

But the reasons I have so far given for believing that the ordinary sentences "John is running" and "The tallest man is running" indirectly ascribe to presentness a monadic property are not sufficient to justify this thesis fully, if only for the reason that I have been using or presupposing several undefined terms or unsubstantiated theses. For example, I have not yet defined "quasi-corresponds," "corresponds," or "state of affairs" and have not yet shown how complexes such as *inhering in John's running* can be regarded as properties. These matters are best handled if we turn our attention from propositions to their corresponding states of affairs. A discussion of states of affairs will also put us in a position to show that presentness is not merely a logical subject of every present tensed A-sentence but is also a metaphysical subject of every state of affairs that corresponds to a true token of such a sentence. Once this is shown, we shall have reached our preliminary goal of showing that presentness functions as a logico-metaphysical subject relatively to all present tensed A-sentences. I will then argue, in the last two sections of the chapter, that presentness is also a logical subject of all past and future tensed A-sentences and that this result is not vitiated by arguments based on McTaggart's paradox.

5.6 States of Affairs

A state of affairs is whatever corresponds to a true proposition. The n-adic property of being true possessed by a proposition is precisely its relation to a state of affairs of corresponding to it. A true proposition is what *originally* corresponds to a state of affairs; a sentence-token, by virtue of expressing a true proposition, corresponds derivatively to the state of affairs. The complex relation of correspondence is analyzable in terms of three simpler components, *identity, replacement,* and *instantiation* such that these latter connections obtain between constituents of the proposition (or, derivatively, constituents of the sentence-token) and constituents of the state of affairs.

Suppose that a proposition expressed by a token of "John is running" is true. In that case, the connections that obtain between these propositional constituents and constituents of the state of affairs are those shown in Figure 5.8, where (P1) includes constituents of the proposition and (S1) constituents of the state of affairs that corresponds to the proposition if the latter is true. There is a relation of *identity,* symbolized by the two straight lines, between John and John and between running and running. But there is a relation of *replacement,* symbolized by the two broken lines, between *( ) is propositionally related to ( ) and ( ) exemplifies ( ). ( ) exemplifies ( ) replaces ( ) is propositionally related to ( )* in the sense that the former *connects* John to running in the proposition and thereby is the *order* among these two parts in the proposition, whereas this ordering function is taken over by *( ) exemplifies ( )* in the state of affairs, such that John and running are not propositionally connected in the state of affairs but are exemplificationally connected, that is, connected through one's *exemplifying* the other. John and running are parts of the state of affairs but *exemplifies,* qua *order* among these parts, is not a part but merely a constituent.

The proposition expressed by a token of "John is running" also includes these propositional constituents, which sustain the connections shown in Figure 5.9. The brackets in (P2) are filled in by (P1) and the brackets in (S2) are filled in by (S1). Here the property tie *replaces* the propositional relation and presentness stands in the relation of *identity* to presentness.

In section 5.4, in connection with my discussion of satisfaction and cosatisfaction, I said that what fills in the brackets in (P2), namely (P1), is satisfied by what fills in the brackets in (S2), namely (S1), in the sense that it *quasi-corresponds* to it. A propositional complex such as (P1) quasi-corresponds to an event, such as (S1), in the sense that the constituents of (P1) are connected to the constituents of (S1) via identity, replacement, or instantiation. In short, quasi correspondence is just correspondence except that the relational terms of quasi correspondence are not complete propositions and states of affairs but propositional complexes and events, which are complex constituents of propositions and states of affairs. A propositional complex is a propositional relatedness of a property to something, and an event is an exemplification of a property by something. (P1) is a complex constituent of the proposition expressed by a token of "John is running," the other complex constituent being (P2). (S1) is a complex constituent of the state of affairs that corresponds to this constituent, the other complex constituent being (S2).

(S1) and (S2) are one way of stating the constituents of the state of affairs S. There is another way. If John is ordered to running through the property tie of

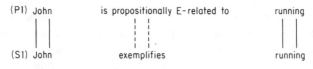

Figure 5.8

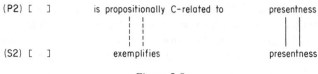

Figure 5.9

exemplifying running, that is tantamount to saying that running is ordered to John through the property tie of *being exemplified by* (inhering in) John. But the property tie of *exemplifying* and the property tie of *being exemplified by* are not identically one and the same tie but are converse ties that together comprise the ordering of a property and an individual (or another property, in the case where one property exemplifies another). The ordering constituent of the state of affairs that replaces the propositional relation is a complex constituent in that it is the ordering of John to running through the tie of *exemplifying* and, conversely, the ordering of running to John through the tie of *being exemplified by (inhering in)*. The same holds for (S2), so that John's running is ordered to presentness through exemplifying it, and presentness is ordered to John's running through inhering in it.

For *de re* propositions such as *that John is running,* the relation of correspondence involves only the relations of identity and replacement. But for *de dicto* propositions such as *that the tallest man is running,* correspondence additionally involves the relation of *instantiation.* The propositional definite description *being the tallest man* is instantiated by a complex constituent of the corresponding state of affairs, namely, by John-as-exemplifying-the-property-of-*being-the-tallest-man* (supposing that John is, in fact, the tallest man). The instantiation relation is depicted by the ladder in Figure 5.10, where (S3) instantiates (P3) in that (S3) includes a part, John, that *exemplifies* the complex property (P3) and another part, the property of *being the tallest man,* that is identical with the property (P3). This explicates what it means to say that "John is the referent of the sense of 'the tallest man'."

These elucidations of the correspondence and quasi correspondence relations put us in a position to address the issue of whether complexes such as *inhering in John's running* are monadic properties of presentness. As argued earlier, a token of "John is running" has the same semantic content as a token of "Presentness inheres in John's running." Both indirectly refer to the complex *being exemplified by John's running* (i.e., *inhering in John's running*). But is this a property? Is it not, rather, a complex of an event (John's running) and a property tie (inherence in)? It seems to me that this question calls for a terminological decision, rather than a "proof" of some negative or affirmative answer; for the following reasons.

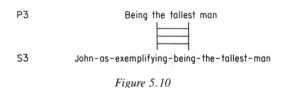

Figure 5.10

Suppose that the standard view that inherence and exemplification are "nonrelational ties" (to borrow Strawson's phrase) is true.[6] This view is defended by Gilbert Ryle, Wilfred Sellars, D. M. Armstrong, Milton Fisk, Alan Donagon, Gustav Bergmann, Irving Copi, Richard Parker, Erik Steinus, Matti Sintonen, William Winslade, P. F. Strawson, and many others.[7] On this standard theory of the property tie, *inherence in John's running* is not a property if "property" is stipulatively defined as

D1 A property = df. A property is whatever can be exemplified by something,

where "something" ranges over concrete and abstract objects. According to D1, *inherence in John's running* is not a property, since it attaches directly or immediately to presentness, rather than mediately through the linkage of a property tie of *exemplification*. This can be stated in another way if we use the word "possesses" not as a synonym or species of "exemplifies" but as a genus of which "exemplifies" is one species. In terms of the example of John's exemplification of running, we may say that on the nonrelational tie theory, John *indirectly possesses* running through *exemplifying* it but *directly possesses* exemplifies-running, since the complex exemplifies-running is "linked to John without an intermediary link." In short, John indirectly possesses *running* through directly possessing *exemplifies-running*.

But suppose that inherence and exemplification are not nonrelational ties but are relations, as is explained by the minority tradition of Meinong, the early Russell, and Nicholas Wolterstorff.[8] Then they are properties (polyadic properties) in the sense (D1). According to this tradition, a benign infinite regress ensues. For example, if presentness exemplifies the property *inheres in John's running*, then presentness also exemplifies the property

(F) exemplifies inherence in John's running

and, by virtue of exemplifying F, exemplifies

(F′) exemplifies exemplifies-inherence in John's running.

If we adopt definition D1 of a property, then *inherence in John's running* counts as a property only if the benign infinite regress theory is true. But this does not force the presentist to accept the benign infinite regress theory; for the presentist could reject the definition D1 and offer a more latitudinarian definition of a property, namely,

D2 A property = df. A property is whatever can be possessed by something, whether indirectly or directly.

Definition D2 allows as properties such items as *wisdom, exemplifies wisdom,* and *inheres in Socrates* but excludes such items as Socrates, Greece, the number two, propositions, events, states of affairs, arguments, sentence-tokens, sets, and the like. "Property" in D2 is meant as "n-adic property," so that it includes both relations like *loves* and nonrelational ties like *exemplifies*.

Suppose, however, that someone accepts as true the "nonrelational tie" theory but nonetheless insists that D2 is false and D1 is true. I would respond that *if* the "nonrelational tie" theory is true, then D1 and D2 do not have truth values, since they are stipulative definitions about how "property" is to be used. If this theory is true, then there are both referents of the term *property* as used in D1 and referents of the term *property* as used in D2, and the issue is merely of which usage of "property" is to be adopted. Accordingly, the presentist who adopted the stipulation D2 would merely differ terminologically from one who chose the stipulation D1.

Now it is not necessary for the purposes of establishing presentism that I decide between the "nonrelational tie" and the "relational tie" conceptions of exemplification and inherence; for presentism is consistent with both conceptions, although the formulation of this theory would vary depending upon which conception is adopted. If the nonrelational tie conception is adopted, then the presentist who wants to call *inheres in John's running* a property will adopt the definition D2 of a property. If the relational tie conception is adopted, then the definition D1 must be adopted, instead; for on the relational tie theory there is no such thing as a directly possessed property. However, to avoid persistent ambiguity in my subsequent formulation of presentism, I shall develop presentism in terms of one of these views, namely, the majority or standard view that inherence/exemplification ties are nonrelational ties and leave it to the interested reader to make the appropriate modifications at the appropriate places if he or she wishes to view presentism through the lenses of the "relational tie" theory. Accordingly, I shall adopt the stipulation D2.

Consider, then, the property *inheres in John's running*. This is a monadic property that is built up out of an event and a property tie. As such, it is analogous to certain *relational properties,* that is, to monadic properties that are built up out of relations and individuals (first-order events or things). Examples of such relational properties are *being earlier than John's running* and *loves Alice.* Such properties exist only if their constituent individuals exist. If there is no event of John running, there is no property of *inhering in John's running* or *being earlier than John's running.* The grammatical predicates "inheres in John's running" and "is earlier than John's running" would have a sense but no referent. (But they would lack even a sense if John did not exist, since these senses include John himself, as explained in the last section.)

Given that complexes of the form *inheres in B,* where B is some being in which presentness inheres, are directly possessed properties, we can understand how the results achieved so far entail that presentness is a *metaphysical subject* of the states of affairs that correspond to true tokens of present tensed A-sentences. Since presentness is a direct referent of the present tense of these tokens, it is a *part* of the corresponding states of affairs. And since presentness is indirectly ascribed a monadic property of the form *inheres in B* by these tokens, presentness belongs to the corresponding states of affairs as *directly possessing* a property of this form. This entails that presentness is a metaphysical subject of these states of affairs, for "x is a metaphysical subject of the state of affairs S" entails "x is a part of S and has the role or place in S of a property possessor (i.e., x belongs to S as possessing, directly or indirectly, some property)." But this does not entail, of course, that presentness is the only metaphysical subject of these states of affairs. If presentness inheres in some being B, then B is also a metaphysical subject of that state of affairs; for B is both a part of that state of affairs and has the role in it of *indirectly possessing the property of presentness,* that is, of exemplifying presentness. In the state of affairs corresponding to a token of "John is running," the event of John's running exemplifies presentness and therefore is a metaphysical subject of this state of affairs. But so are John and running, since John belongs to the state of affairs as exemplifying running and running belongs to it as directly possessing *inhering in John.* States of affairs contain a multiplicity of metaphysical subjects. But if all states of affairs include presentness as a metaphysical subject and there is nothing else that is a subject of every state of affairs, then presentism is true.

I shall argue in chapter 6 that nothing besides presentness is a subject of every state of affairs. But before further arguments about the parts of states of affairs are developed, it is necessary to achieve a greater clarity about the notion of a state of affairs itself. Specifically, it is necessary to distinguish my usage of "state of affairs" from the usage of this expression by many other contemporary philosophers, since a failure to distinguish these two uses will inevitably result in a misunderstanding of the theory of states of affairs that I am developing. Many philosophers use "state of affairs" to denote a *truth-valued complex,* rather than a *truth-maker* of such complexes, whereas states of affairs in my sense are what make any truth-valued complex true. Philosophers such as Alvin Plantinga, Roderick Chisholm, and Nicholas Wolterstorff use "state of affairs" to denote truth-valued complexes.[9] My difference from Chisholm and Wolterstorff can be easily stated, since they use "state of affairs" to refer to what I call "propositions." They mean by "a state of affairs obtains" what I mean by "a proposition is true," and they mean by "a state of affairs does not obtain" what I mean by "a proposition is false."

But matters are less clear when we consider Plantinga's ideas. According to Plantinga, states of affairs are neither propositions nor truth-makers of propositions. They are not truth-makers of propositions because, if a certain state of affairs exists, that does not suffice to make the relevant proposition true. In order for the relevant proposition to be true, the state of affairs must not only *exist* but also *obtain;* and the obtaining of the state of affairs is itself due to certain objects exemplifying certain properties. It seems to me, however, that there are no "states of affairs," if this expression is used in Plantinga's sense to refer to something different from both propositions and the truth-makers of propositions. A consideration of this point should help clarify the import of my statement, "I use 'states of affairs' to refer to truth-makers and not to truth-valued complexes."

Plantinga says that "there are such things as states of affairs; among them we find some that obtain, or are actual, and some that do not obtain, . . . [an example of the latter being] *Spiro Agnew's being President of Yale University.*"[10] Furthermore, Plantinga continues, there is an intimate relation between states of affairs and propositions. If there is a state of affairs *Spiro Agnew's being President of Yale University* that does not obtain, then there is a related false proposition *that Spiro Agnew is President of Yale University;* and if there is an obtaining state of affairs *Socrate's being snub-nosed,* then there is a related true proposition *that Socrates is snub-nosed.* Plantinga writes of this intimate relation between states of affairs and propositions; "Roderick Chisholm, indeed, thinks the relation so intimate as to constitute identity. As he sees it, there are not *two* kinds of entities—propositions and states of affairs—but only one; propositions just *are* states of affairs."[11] Although Plantinga does not commit himself to this identification and continues to talk as if they are distinct, I believe that this identification must be made *if* "states of affairs" is not being used in my sense to refer to the truth-makers of propositions. A state of affairs S is a truth-maker of a proposition P if the existence of S suffices to make P true. More exactly, S is a truth-maker of P if and only if (a) P's being true consists in its correspondence to S and (b) S's existence is both sufficient and necessary for the relation of correspondence to obtain between P and S. But Plantinga denies this of his states of affairs; for example, he writes that "*there are* such things as states of affairs . . . and some of them do not obtain . . . [an example being] *Spiro Agnew's being President of Yale University*." Now if the existence of this state of affairs is consistent with the falsity of the proposition *that Spiro Agnew is President of Yale University*, then the existence of this state of affairs is not what makes the proposition true. Rather, what makes it true (if anything does) must be something other than the state of affairs, namely, Spiro Agnew's exemplification of the property of *being President of Yale University*. If this truth-maker exists, that is, if Agnew exemplifies this property, then the proposition is true. And if this truth-maker exists, then Plantinga's state of affairs *Spiro Agnew's being President of Yale University* not merely exists but also *obtains*.

But what are these strange "states of affairs" of Plantinga? Plantinga distinguishes between their existence and their obtaining, since some states of affairs exist but do not obtain. This means they are, in a relevant sense, truth-valued complexes, since their "obtaining" and "not obtaining" is exactly analogous to the "being true" or "being false" of propositions. It also means they require truth-makers in order to be true (to "obtain"). But it is hard to see, then, how Plantinga's states of affairs differ, except verbally, from his propositions. This problem is exacerbated by the fact that the *constituents* of Plantinga's states of affairs seem indiscernible from the constituents of his propositions. Consider that the proposition expressed by a *de dicto* use of the sentence "The tallest human is wise" consists of the properties of being the tallest human, the property of wisdom, and a temporal property. But Plantinga's state of affairs *the tallest human's being wise* also appears to consist of all and only these properties (as ordered to each other in the same way) and thus appears to be identical with the proposition. If there is some difference between the constituents of the state of affairs and the constituents of the proposition, Plantinga does not tell us what it is. To say that one "obtains or does not obtain" whereas the other "is true or false" seems to be merely a verbal difference, given the absence of any difference in the constituents of the two complexes.

In summary, I differ from Chisholm and Wolterstorff in that I use "propositions" to denote what they denote by "states of affairs" and differ from Plantinga by denying that there is anything at all denoted by his use of "state of affairs" (unless this is just a different name for his propositions). In both cases, my use of "states of affairs" is different, since I use this phrase to denote the truth-makers of propositions. Given that states of affairs are truth-makers, it makes no sense to distinguish between the existence and obtaining of a state of affairs, since it is senseless to talk about a truth-maker's existing but not obtaining. This makes sense only if states of affairs are not truth-makers but truth-valued complexes (specifically, propositions) and "exists but does not obtain" just means "exists but is false." This follows from the theory of propositions and states of affairs outlined in this chapter. Consider the true *de re* proposition *that John is running* and its corresponding state of affairs. Both the proposition and the state of affairs have for their parts John, running, and presentness. But they are ordered in the proposition in a different way from how they are ordered in the state of affairs. They are related in the proposition in such a way that (1) the existents can be so related even if John is not presently walking (but given that the proposition is *de re,* they cannot be so related if John does not exist, since John is a part of this proposition only if he exists); (2) the complex consisting of the existents as so related has a truth value; and (3) the complex is not identical with John's present walking (this is entailed by condition [1]). But John, walking, and presentness are related in the state of affairs in such a way that (1) they are so related if and only if John is presently walking; (2) their being related in this way does not possess a truth value but corresponds to something (namely, the proposition) that possesses the value of true; and (3) the state of affairs is identical with John's present walking (i.e., it is not a set or merelogical sum or aggregate composed of John's present walking, it just *is* John's present walking and is nothing different from this or over and above this). The state of affairs is John's *exemplifying* walking and his exemplifying walking's *exemplifying* presentness. This entails that if *there is* the state of affairs, then John *is* presently walking; accordingly, it makes no sense to say that there is this state of affairs but that it does not obtain, since this could only mean "There is this state of affairs, but John *is not* presently walking," which, given my definition of a state of affairs, is an implicit logical contradiction. Accordingly, if states of affairs are truth-makers, then there is no distinction between their existence and their obtaining, such that "the state of affairs S exists" means "S obtains" and "S does not obtain" means "S does not exist."

What is missing from the philosophy of Chisholm, Wolterstorff, Plantinga, and others is the notion of a truth-maker of a proposition, that is, the notion that I (but not they) express by "state of affairs." They would not deny that there is something that makes a proposition true, but they have no theory of such items; and in this respect, their philosophy is impoverished.

So far in this chapter, I have explained the nature of propositions and states of affairs, with the end in view of establishing that presentness is a metaphysical subject of each state of affairs that corresponds to a proposition expressed by a present tensed A-sentence. The remaining task of this chapter is to show how the results achieved so far enable us to show that (a) presentness is also a metaphysical subject of the states of affairs that correspond to past and future tensed A-sentences (see section 5.7) and (b) McTaggart's paradox does not vitiate the theory that presentness is a logico-metaphysical subject (see section 5.8).

5.7 Presentness as a Metaphysical Subject of All A-States-of-Affairs

An A-state-of-affairs is any state of affairs that corresponds to a proposition expressible by an A-sentence-token, be it present tensed, past tensed, or future tensed. I argued in the last section that presentness was a metaphysical subject of every state of affairs that corresponds to a proposition expressible by a true *present* tensed A-sentence token. But is presentness also a metaphysical subject of all pastness-and-futurity-involving states of affairs? It is such a subject if all true past and future tensed A-sentence-tokens ascribe a complex property—present pastness or present futurity—to which presentness belongs, as I shall now argue.

Let us suppose that "John was running" does not ascribe present pastness but merely *simple pastness,* where "simple pastness" designates a property of which presentness is not a part. In that case, "John was running" would be synonymous not with "John's running is past" (where the copula is present tensed and thereby indicates that John's running is *presently past*) but with "John's running (is) past," where the copula is tenseless. The latter sentence does not indicate *when* John's running exemplified simple pastness. It does not tell us whether John *will* exemplify it, *is* exemplifying it, or *was* exemplifying it. "John's running (is) past" is true if and only if John's running exemplifies pastness *at some time,* be this time now future, present, or past. This shows that "John was running" does not have the same semantic content as "John's running (is) past" and therefore does not ascribe merely simple pastness to John's running. "John was running" has the semantic content of "John's running is past," which tells us that John's running *presently* exemplifies pastness, that is, that it has *present pastness.*

But exactly what is the complex property of present pastness? It cannot be the *conjunction* of simple pastness and presentness, since "John was running" does not mean that the property of *being present and past* is exemplified by John's running. Nor can it be the property of pastness as exemplifying a second-order property of being present; for "John was running" does not mean "John's running exemplifies the property of pastness, and the property of pastness has the property of being present." Apart from the difficulty of saying that the property of pastness is itself something that occupies the present, it does not avoid the aforementioned problem of indicating when John's running has pastness; for even if pastness is itself present, that does not tell us whether this property is presently *exemplified* by John's running or whether it *will be* exemplified by his running. The complex property of present pastness, inasmuch as its possession by something tells us "when that item has pastness," must be the property consisting of the *inherence of presentness in the inherence of pastness in ( ),* where the brackets are filled in by the past item. The possession of present pastness by John's running is such that the inherence of pastness in John's running is something that obtains at present. The state, the exemplification of pastness by John's running, is something that exemplifies presentness.

Accordingly, to ascribe present pastness to John's running is to ascribe simple pastness to John's running and is to ascribe presentness to the inherence of simple pastness in John's running. We have these propositional relations:

(P1) Simple pastness   is C-related to   John's running.

(P2) Presentness   is C-related to   the pastness of John's running.

"John's running" in (P1) means John-as-E-related-to-running and "the pastness of John's running" in (P2) just means (P1), that is, simple pastness as C-related to

John's running. If the proposition is true, then presentness is exemplified by what satisfies (P1), namely, the state or event, *the inherence of simple pastness in John's running*. (I use the words *event* and *state* as synonyms to designate the exemplification of a property by something, as I shall formally define in chapter 6.) The corresponding state of affairs consists of presentness as directly possessing the property *inhering in the simple pastness of John's running*. This entails that presentness is a metaphysical subject of this state of affairs.

Similar results hold for true future tensed A-sentence-tokens. If a token of "John will run" is now true, it now corresponds to a state of affairs consisting of presentness as inhering in the inherence of simple futurity in John's running. But I would point out that "Presentness is a metaphysical subject of all future states of affairs" is consistent with the position that "A-sentence-tokens about future contingents are now neither true nor false." It sometimes argued that tokens of sentences such as "John will run" are not now true or false, since that would entail fatalism. I shall not decide about this issue here but simply point out that my theory is that for any true future tensed sentence-token, the future state of affairs to which it corresponds includes presentness as a metaphysical subject. If tokens of "John will run" now possess no truth values, they now correspond to no states of affairs, which is consistent with the fact that presentness is a subject of any future states of affairs that in fact obtain. Furthermore, the fact that sentence-tokens about future contingents have no truth value is consistent with there being some future tensed sentence-tokens that have truth values; for the truth-valuelessness of a token of "John will run tomorrow" is consistent with the truth of "Tomorrow will become present."

I will add that if the propositions expressed by sentence-tokens about future contingents are truth-valueless, that may require some tinkering with the earlier definition of propositions as original truth-value bearers; but the modifications that may be required are easily made and I shall not pause here to develop them. (An even simpler move, requiring no modifications, would be to classify the semantic contents of truth-valueless sentence-tokens as "pseudopropositions."

It is of more importance to examine the following criticism of the theory presented in this section. It may be objected that there are no past or future states of affairs but only present states of affairs. There are no past or future states of affairs, the objection goes, since that would require nonexistents to presently possess properties, namely, properties of pastness and futurity; and it is logically impossible for nonexistents to presently possess properties. This is a position held by A. N. Prior, Ferrel Christensen, Genevieve Lloyd, A. B. Levison, William Lane Craig, and many others.[12] It seems to me, however, that the argument for this thesis commits the informal fallacy of question-begging.

A. N. Prior writes that "getting more and more past seems to be something an event does when it *doesn't* exist, and this seems very queer indeed";[13] but he leaves it to others to construct an argument that events cannot possess a property of pastness (or futurity). Ferrel Christensen musters the following argument:

> In order to have any properties at all, as that term is normally understood, an entity must exist; to have a property now it must exist now, to have had a property earlier it must have existed then. Hence it would also seem that to have such properties as

presentness and pastness, an event or state of affairs would have to exist. That is, it exists for an indefinitely long period during which it has the property of futurity, it then momentarily has the property of presentness, and it has the property of pastness, while continuing to exist, ever after. . . . Such an interpretation is false: to say that an event is past and no longer present really only means that it *did* exist but *no longer* exists, that it *was* occurring but is *not*.[14]

This suggests the following argument:

(1) Something now possesses properties if and only if it now exists.

(2) Something now exists if and only if it is present.

Therefore,

(3) Something now possesses properties if and only if it is present.

Item (3) entails that pastness and futurity are not properties, since if they were properties, they would be now possessed by items that are not present, which contradicts (3). But this argument is unsuccessful, since either it is question-begging, or its premises are mutually inconsistent. If the premises are consistent, the argument is question-begging, since it assumes what it wants to prove, namely, that items now possess properties if and only if they are present. In this case, "exists" in (1) either means or is logically equivalent to "is present"; thus, (1) simply assumes the falsity of the view that pastness and futurity are properties now possessed by items that are not present. But if the argument is rendered non-question-begging, the premises become inconsistent. It is non-question-begging if "exists" in (1) is used tenselessly, to mean or to be logically equivalent to, "is past, present, or future"; but (1) is then inconsistent with (2).

Question-begging arguments of this sort are a commonplace in the literature on the tensed theory of time. Another example can be found in C.J.F. Williams's *What is Existence*? He writes:

We cannot say of those who no longer exist that they are at this moment engaged in any activity. The dead are not now up and doing. How much is included in the scope of "activity" and "doing" here? There is a sense of "do" in which it is a mere dummy verb, a predicative variable, as "thing" is a dummy noun or individual variable. In this sense of "do" we might enunciate as a general law that if *a* is truly said to do anything at time *t* there must be such a thing as *a* at time t. Meta-linguistically this can be expressed thus: if any predicable is truly applicable to (holds of) an object at a given time there must at that time be such an object.[15]

Williams begins with the plausible premise that those who no longer exist are not now engaged in any activity, they are now not up and doing. Socrates is not now discoursing on the streets of Athens. But Williams, by equivocating upon "do," passes from this plausible premise to the question-begging premise that predicates apply to something at a given time only if that item occupies that time. He extends "do" from its normal sense of referring to activities like talking, walking, and the like to his technical sense where it denotes any property, anything expressible by a predicate. But the general law he enunciates does not follow from the plausible premise about "activities" in the normal and restricted sense of the term and represents a mere assumption of the falsity of the view that items that no longer exist now possess some properties.

Michael Dummett makes the related claim that items cannot possess properties of no longer being present or no longer existing. He claims that "existence, even when temporal, is not a property that may first be acquired and later lost"[16] and justifies this claim by the argument that there is an "absurdity in saying that there is such a person as Cleopatra, who no longer has the property of existing."[17] But this is absurd only if "there is" is present tensed and means "there now exists." If it is tenseless and means "there was, is now, or will be" and "existing" is present tensed and means (or is logically equivalent to) "is present," then the statement is not absurd, since it is equivalent to "there was, is now, or will be such a person as Cleopatra, who no longer has the property of being present." Unless there is some further argument that shows "there is" cannot be assigned this tenseless meaning, Dummett's claim is unjustified. Since Dummett supplies no such argument, there is no reason to accept his claim.

A stronger argument against proponents of the thesis that "only what is present can possess a property" can also be developed, namely, that their thesis is not only unjustified (being based on question-begging or unsound arguments) but is also inconsistent with another position they also hold, namely, that *some tokens of past or future tensed A-sentences are true*. These tokens are true, I shall argue, only if some past or future events or substances presently exemplify some properties.

We may begin by considering Wolterstorff's theory. Wolterstorff holds the view that "something can presently have a property only if it presently exists."[18] Wolterstorff at least admits that this is a mere assumption and constructs no question-begging argument for it. He says, "[it] does in fact seem to me to be true. I shall here not argue the case, however."[19] But Wolterstorff's assumption is inconsistent with his position that some tokens of past tensed A-sentences are true. Wolterstorff considers his past tensed proposition (8):

(8) Bucephalus crossed the Meander

and writes:

> It cannot be that this is true just in case Bucephalus presently has the property of having-crossed-the-Meander. For (8) is presently true, yet Bucephalus presently does not exist. And by Ontological Principle II, something can presently have a property only if it presently exists. (8) must rather be construed as identical with this:
>
> > (9) There was something such that it had the property of being-identical-with-Bucephalus and simultaneously had the property of crossing-the-Meander.[20]

Wolterstorff makes no attempt to explain the semantic content of (9). This is just as well, since such an attempt would result in the realization that (9), if true, is inconsistent with his Ontological Principle ("Something can presently have a property only if it presently exists"). Let us begin by asking about the semantic content of "something" in (9). Wolterstorff seems to be working with objectual, rather than substitutional, quantification, so "something" as used in (9) may be understood as containing "thing" as a first-order variable whose values are individual things— substances like horses and persons and stones. Suppose these substances are present. Then (9) is false, since there is no present substance (horse, person, stone, etc.) that had the property of being identical with Bucephalus and of simultaneously crossing the Meander. But suppose "something" ranges over past substances. Then (9) is true but conflicts with Wolterstorff's Ontological Principle, since these past substances presently possess the relational property of *being values of the variable "thing" in (9).* These past substances are now being denoted by this variable and— as now standing in this relation to it—now possess some dyadic property. Moreover, these substances are now being mentally referred to by whoever is uttering the sentence (9) and thus now exemplify *being mentally referred to.* Furthermore, since they *are now past,* they presently possess the property of pastness.

An analysis of (9) that makes it come out true also implies that there are past events that presently possess properties. According to (9), something "had the property" of crossing-the-Meander (which Wolterstorff regards as preferable to "something now has the property of having-crossed-the-Meander"). But what does "had the property" mean? Wolterstorff makes no attempt to explain its meaning. But if (9) is true, it must mean that something's *having* (possessing) the property of crossing-the-Meander is now past. However, the possession of a property by something is an event. This implies that an event, something's possession of the property of crossing-the-Meander, is now past. This past event now possess properties, such as being past, being referred to, and so on.

I would add that if we assume that this event is past, we can deduce from this that a certain substance, namely, Bucephalus, is also past. The facts that (i) Bucephalus is not present or future and (ii) his exemplification of crossing-the-Meander is now past jointly entail that (iii) Bucephalus is now past. It is an implicit contradiction to assert that Bucephalus is neither past, present, nor future but that his crossing the Meander is past; for this would mean either (a) Bucephalus exists timelessly and yet a few centuries ago crossed the Meander or (b) Bucephalus never exists and yet once crossed the Meander.

It is possible that this analysis of (9) may be rejected. The variable "something" may not be interpreted as a first-level referring expression (that denotes substances) but as a second-level predicable. Suppose that "something" in (9) is interpreted as expressing the property of being exemplified that is possessed by being-Bucephalus and crossing-the-Meander. "There was something such that it had the property of being Bucephalus" would mean (or at least entail) "Being-Bucephalus *was exemplified.*" Since being-Bucephalus is a first-order property, it was exemplified by a first-order existent, in this case a substance. If (9) is true, being-Bucephalus was exemplified by Bucephalus, such that being-Bucephalus stands in the property tie (or relation, if you prefer) to Bucephalus of having been exemplified by him. Note that "being exemplified" is not the most perspicuous way of expressing this property, since *being exemplified* is not a monadic property but a dyadic property, namely, the property ( ) *being exemplified by* ( ). Accordingly, two terms are required if this property is to be possessed. In our case, the two terms are being-Bucephalus and

Bucephalus himself. Since we are dealing with a past exemplification of being-Bucephalus, with the dyadic property ( ) *having been exemplified by* ( ), one of these terms (i.e., Bucephalus) must be past. This entails that Bucephalus now possesses the property of *having exemplified being-Bucephalus*. Since Bucephalus is now past, this entails that a past substance presently possesses some property, namely, *having exemplified being-Bucephalus*. If Bucephalus does not now exemplify this property, then (9) is now false; for then there would be nothing (past, present, or future) that is now connected to the property of *being-Bucephalus* by the relation or property tie of ( ) *having exemplified (it)*.

If "Being-Bucephalus was exemplified" is now true, that also entails that some event or state is now past. If this sentence is true, then *being exemplified by* ( ) is not now possessed by the property of being-Bucephalus; rather, it was possessed by this property. This entails that the event, the being exemplified of being-Bucephalus, is past and presently possesses some property, namely, the property of being past.

Similar considerations show that a still further interpretation of (9) fails to support Wolterstorff's contention that only what is present presently possesses properties. Item (9) may be interpreted as meaning

> (10) It was true that something has the property of being-Bucephalus and simultaneously has the property of crossing-the-Meander.

Since the variable is within the scope of the past-tense operator, it does not now range over present or past substances; rather, it used to range over substances—substances that were then present.

This may be granted; but it fails to preserve Wolterstorff's Ontological Principle, since if the variable used to range over substances, then its *ranging over substances* is now past. This implies that some past event or state now possesses some property: the variable's *exemplification of ranging over substances* now possesses the property of pastness. Furthermore, we still have past substances; for if the variable used to range over substances, the substances that were relata of the relation ( ) *ranging over* ( ) can only be those that are either now past or that were then present and still continue to be present. Since Bucephalus is not now present and (10) is now true, he must belong to the class of substances that now exemplify pastness.

Similar results follow from an analysis of (10) as a sentence that ascribes truth to a proposition. We may ask, When does the proposition expressed by "Something has the property of being-Bucephalus and simultaneously has the property of crossing-the-Meander" possess the property of truth ascribed to it? It does not now possess this property. Rather, it used to possess it. This gives us a past event or state, the exemplification of *being true* by the proposition. The proposition's being true *now* exemplifies pastness; since this is a past event or state that presently possesses a property (pastness), Wolterstorff's Ontological Principle is violated. But this is not all. If the proposition's being true now exemplifies pastness, then the proposition's *standing in the relation of correspondence* to a state of affairs now exemplifies pastness. This entails that some state of affairs stood in the relation of correspondence to this proposition, that is, that some state of affairs' *standing in the relation of correspondence* to the proposition now exemplifies pastness. The state of affairs that stood in this relation is Bucephalus-crossing-the-Meander. This state of affairs occupies the very same position in time as does the event of its *standing in the relation of correspondence* to the proposition; that is, it is past. This implies that the state of affairs of Bucephalus-crossing-the-Meander presently possesses the property of pastness.

These several arguments suggest that the thesis that *some past or future tensed A-sentence-tokens are presently true* entails that some substances, events, or states of affairs that are not present *presently* possess some properties.

Does this conclusion entail that "nonexistents possess properties"? That depends on the interpretation given to "nonexistents" and "possess." If "nonexistents" means "what is not present" and "possess" means "presently possess," then some nonexistents possess properties; for past substances and events presently possess some properties. But if "nonexistents" means "what is neither past, nor present, nor future," then nothing I have said entails that nonexistents possess properties. I believe that if something possess a property, then it is past, present, or future. (In Chapter 6 I argue that nothing is timeless.)

Correlatively, we may say that "x exists" in the present tensed sense is not a necessary condition of present property possession but that "x exists" in the tenseless sense is a necessary condition of present property possession. "x exists" in the tenseless sense may be defined in terms of the present tensed sense, so that "x exists" in the tenseless sense means "x existed, exists, or will exist," where the middle "exists" is present tensed. "x existed, exists or will exist" means, or is logically equivalent to, "x is past, present, or future"; and "x exists" in the present tensed sense means, or is logically equivalent to, "x is present."

It is worthwhile to add that the analyses just given show there is a present tensed sense of "exists" in which past and future items can be said to exist, namely, the sense where "x exists" means "x presently possesses some properties." Past and future items exist in this present tensed sense, since they presently possess properties of pastness or futurity. Thus, we may respond to Christensen, Williams, Wolterstorff, and others that even if it is assumed that "x exists"—where "exists" is present-tensed, is a necessary condition of the truth of "x now possesses properties"—it does not follow that past and future events do not now possess properties, since it is true of each past and future x that "x exists" in the present tensed sense just explained.

In the past several pages I have been responding to an objection to my theory of past and future states of affairs based on the allegation that past and future events cannot presently possess any properties and therefore that presentness is not a metaphysical subject of any past or future state of affairs. But there is also a more fundamental objection to presentism, an objection also propounded by Prior, Christensen, Williams, and others. This is the objection that the tenseless theory of time is false but that it is nonetheless the case that *tenses do not ascribe A-properties to events* and therefore that presentness is not a metaphysical subject of the states of affairs that correspond to true tokens of A-sentences. It seems to me, however, that this objection is either unintelligible or self-contradictory, as I shall argue in the next section.

5.8 The No-Property Tensed Theory of Time

My argument that presentness is a metaphysical subject of all states of affairs corresponding to true A-sentence-tokens has involved the thesis

(1) The tenses of true A-sentence-tokens directly ascribe A-properties.

As is well known, (1) is denied not only by detensers but also by many tensers, proponents of (what I shall call) the *no-property* tensed theory of time. These include C. D. Broad, A. N. Prior, F. Christensen, G. Lloyd, A. Levison, Delmas Lewis, Gilbert Plumer, and many others. The no-property theory makes two claims: (i) that tenses do not ascribe A-properties and (ii) that the thesis that they ascribe A-properties leads to McTaggart's paradox. In section 5.9 I shall show that the A-property theory does not lead to McTaggart's paradox. In the present section, I shall show that claim i, as made by the no-property tensers, is either unintelligible or self-contradictory.

Ferrel Christensen makes the claim that tenses

> perform a very different task from telling what the nature or properties of an individual are: they tell *when it has* its properties. That is to say, they tell whether it once did, or does now, or will yet possess such-and-such characteristics (or bear such-and-such relations), and also simply whether it did, does or will exist. . . . There are no such *individuals* as the past, the present and the future, and no such *properties* as pastness, presentness and futurity. . . . [The information conveyed by tenses cannot be represented in] an ontology of individuals and their properties and relations.[21]

We learn from Christensen that the semantic relata of tenses are not properties, relations, or individuals. What are they? If tenses tell us "when individuals have their properties," in what ontological category should we place *what is told*? Christensen does not say. Consider the state of affairs that corresponds to a true token of "John is running." John and running are parts of this state of affairs, but no A-property and no "A-individual" or "A-relation" (whatever these could be) belong to this state of affairs. What belongs to this state of affairs that correlates to the present tense of the "is"? If the present tense tells something, and is not an idle syntactical device that conveys no information about realty, what it tells us presumably belongs to this state of affairs. But what is it? Christensen does not say.[22] His theory is *ineffable;* it tells us what the relata of tenses are not but not what they are. I would go further and say that his theory is *unintelligible,* since an intelligible theory of the semantic relata of tenses either classifies these relata in terms of some familiar category (e.g., properties) or introduces and explains some new and unfamiliar category in terms of which these relata are classified, and Christensen's theory does neither.

The same holds for A. B. Levison's theory; for he claims that "there is no ontological commitment either to temporal objects, properties, or relations in the ordinary uses of such expressions [i.e., A-expressions],"[23] but forgets to tell us to what these expressions do commit us. And consider this related obscurism from Genevieve Lloyd's pen: "Tense is seen not as another property which can be forced, however uncomfortably, into the subject–predicate mould. Tense does not concern what is true of a thing. It concerns *when* things are thus and so."[24]

The person originally responsible for this unintelligible theory is A. N. Prior, upon whom Christensen, Levison, Lloyd, and others lavish praise. These philosophers base themselves largely on the following passage from Prior:

> "Is present," "is past," etc., are only quasi-predicates, and events only quasi-subjects. "X's starting to be Y is past" just means "It has been that X is starting to be Y," and the subject here is not "X's starting to be Y" but X. And in "It will always be that it has been that X is starting to be Y," the subject is still only X; there is just no need at all to think of *another* subject, X's starting to be Y, as momentarily doing something called "being present" and then doing something else called "being past" for much longer; and no need to argue as to whether X's starting to be Y "is" only at the moment when it does the thing called being present, or also throughout the longer period when it does the other thing. It is X which comes to have started to be Y, and it is of X that it comes to be always the case that it once started to be Y; the other entities are superfluous, and we see how to do without them, how to stop treating them as subjects, when we see how to stop treating their temporal qualifications ("past," etc.) as predicates, by rephrasings which replace them with propositional prefixes ("It has been that," etc.) analogous to negation.[25]

Prior is suggesting that there are no events and no A-properties of events. Instead, there are only substances that possess properties other than A-properties. But what are the semantic relata of tenses if not A-properties of events? Prior claims that a sentence such as "John was running" can be rephrased as

(1) It was the case that John is running.

But how does that answer our question about the semantic relata of tenses? Prior claims it shows these relata *are not* A-properties of events. But what does it show they *are*? The semantico-ontological information that Prior gives us is merely negative, and his positive information is merely *syntactical;* that is, he gives us a *syntactical correlate* of "John was running," namely, (1), but does not explain its semantic correlate. Prior develops at great length and detail the syntactics of tense logics but leaves unaddressed such questions as What is the semantic correlate of the present tense of the "is" in (1)? And what is the semantic correlate of the past tense of the "was" in (1)? Of course, Prior does say that the embedded clause in (1) expresses a "proposition" and that the sentential prefix expresses a "propositional operator"; but that does not answer my questions. What I want to know is what part or constituent of the proposition the present tense of the "is" expresses and what part or constituent of the propositional operator the past tense of the "was" expresses. This information is not forthcoming.

I stated at the beginning of this section that the no-property tensed theory of time was either unintelligible or self-contradictory. I have so far suggested that it is unintelligible. I now want to show that insofar as it can be given an intelligible interpretation (i.e., an interpretation in terms of a familiar category), it is self-contradictory. D. H. Mellor understands the position of Prior, Christensen, and others to be that A-properties are properties of propositions, rather than events. Using "P," "N," and "F" to stand for pastness, presentness (nowness), and futurity, Mellor writes:

> Tense logicians mostly prefer to treat "P," "N" and "F" as "operators" (analogous to "It is not the case that" or "It may be the case that") prefixed to present tense core sentences or propositions. This is tantamount to regarding P, N and F as properties, not of events, but of tensed facts. Where, for example, McTaggart and I start with a thunderstorm, tense logic starts with the sentence or proposition saying that a thunderstorm is happening now. Where we say the thunderstorm is two days past, they say the fact of its happening now is two days past, i.e., the present tense sentence or proposition saying that it is happening now was true two days ago.[26]

Is this how the theory of Prior, Christensen, and company is to be understood? Nowhere do they state that there are A-properties of tensed facts or propositional truth values. But occasionally, they say things that might seem to suggest such a view. For example, Christensen remarks:

> That existence, and hence also non-existence, are not qualities which an individual might equally well have or lack among other properties, is commonly acknowledged. And we can now see that this is one part of the reason why pastness and presentness are not real properties. (We might wish to put it by saying that the tense-operator [e.g., "It was the case that"] and existential quantifier ["There exists"] are together converted into the predicate [e.g., "is past"].[27]

This passage does not wear its meaning on its sleeve, but Christensen seems to be alluding to the Fregean–Russellian theory that "existence is not a property." But what is meant by this clause is not that *existence is not a property* but that *existence is not a first-order property*. Existence is a property of concepts, their instantiation (Frege), or a property of propositional functions, their being sometimes true (Russell).[28] Could it be that Christensen is obscuring alluding to the thesis that A-properties are second-order, rather than first-order, properties? Could this also be the meaning behind Ernest Sosa's obscure statement that "a tensed copula goes with tensed quantification and leaves no room for transitory temporal properties such as being present, being past, being future, etc.,"[29] that is, that *being present* is not a first-order property but a second-order property perspicuously ascribed in tensed quantification sentences? And could this be the meaning of Prior's dark saying, "To be (or become) is simply to happen. It is a kind of zero tense-inflexion; the presentness of a happening is simply its happening."[30] If "happening" or "occurring" is an "existence-verb," as Christensen says,[31] then Prior's saying may be interpreted as meaning that presentness, like existence, is a second-order property.

Now it seems to me that the writings of these philosophers are too obscure to decide with certainty whether they did, in fact, believe that A-properties are second-order. But it is worthwhile, nevertheless, to examine this view on its own grounds. I propose to show that each of the following two theses is self-contradictory:

(1) A-properties are ascribed by tensed quantification sentences but are not first-order properties.

(2) A-properties are ascribed to the truth values of propositions but are not first-order properties.

A first-order n-adic property is an n-adic property possessed by a concrete item (i.e., a substance or concrete event), where a concrete event (or "concrete state") is the exemplification of an n-adic property by a substance (or substances), for example, the exemplification of *exploding* by a furnace or the exemplification of *self-identity* by a person. Consider the present tensed quantification sentence "There is an explosion." This sentence ascribes the second-order property of *being exemplified* to the property of *exploding*. But it ascribes presentness not to the property of exploding but to the *being exemplified* of this property. It states not that the property of exploding is present but that this property is presently exemplified. But this means that the property of presentness is a first-order property; for the property of exploding is exemplified by a substance (e.g., a furnace), and the exemplification of properties by substances are concrete events. Accordingly, (1) is self-contradictory.

Consider now the *de re* proposition *that John is running*. Suppose we say that this proposition is presently true, that is, that presentness, rather than pastness or futurity, inheres in the correspondence of this proposition to reality. If this proposition presently corresponds to a state of affairs, then a state of affairs presently stands in a correspondence relation to the proposition. Now a correspondence relation is a complex relation involving relations of the constituents of the state of affairs to the constituents of the proposition. If a state of affairs presently corresponds to a proposition, then constituents of the states of affairs presently stand in relations to the constituents of the proposition. Since relations are polyadic properties, this implies that constituents of the state of affairs presently exemplify these polyadic properties. Now some constituents of these states of affairs are substances. John is a constituent of the state of affairs corresponding to *that John is running* and presently stands in a correspondence relation of *being identical with* to a part of the proposition, namely, John. Since John is a substance, his exemplification of the dyadic property of *being identical with* is a concrete event or state; and since presentness inheres in his exemplification of this dyadic property, it follows that presentness is a first-order property.

I conclude, then, that the no-property tensed theory of time is either unintelligible or self-contradictory. It remains to show that the thesis *A-properties are first-order properties* does not lead to McTaggart's paradox.

5.9 McTaggart's Paradox

I shall first present McTaggart's paradox, expose its flaw, and then examine the versions of his paradox that are developed and defended by Oaklander, Dummett, Mellor, and Schlesinger.

The basic premise of McTaggart's paradox is that the tensed theory of time entails that for each event E, it is true that

(1) Event E is past, present, and future.

According to McTaggart, (1) is self-contradictory, since the three properties are incompatible. Note that this evaluation requires (1) to be understood as predicating the three properties nonsuccessively, since only in this case would (1) be self-contradictory. This is admitted by McTaggart: "The attribution of the characteristics past, present and future to the terms of any series leads to a contradiction, *unless it is specified that they have them successively*."[32] But there is some ambiguity about (1)'s nonsuccessive predication of the three properties. If (1) predicates the three properties nonsuccessively, then it predicates them either *simultaneously* or *timelessly*. If (1) predicates them simultaneously, it is self-contradictory, since E cannot be past, present, and future at one and the same time. If (1) attributes the three properties to E both nonsuccessively and nonsimultaneously, then it attributes them to E timelessly, which is self-contradictory, since E cannot be timelessly past, present, and future. McTaggart is not perfectly clear as to which of these two interpretations of (1) he adopts. According to some commentators, such as Pavel Tichy,[33] McTaggart means that E possesses the A-properties timelessly, whereas according to others, such as Oaklander,[34] McTaggart's theory is that E possesses them simultaneously. I will leave this issue undecided, since it will not affect my argument. It is also not clear whether the "is" in (1) is tenseless or present-tensed. Broad, Prior, Christensen, and most other commentators[35] believe McTaggart held it to be tenseless; but Oaklander[36] believes McTaggart held it to be present-tensed. I shall also not decide this issue, since my argument against McTaggart remains sound on either interpretation. It suffices to say that McTaggart regards (1) as self-contradictory because it predicates (in a tenseless or present tensed manner) the three properties *nonsuccessively* of E.

McTaggart proceeds to argue that the contradiction in (1) cannot be eliminated by moving to a higher level at which the properties are possessed successively. The expansion of (1) into

(2) E is past at a future moment, present at a present moment, and future at a past moment

is unsuccessful because it is true of each of these moments M that

(3) Moment M is past, present, and future.

Item (3) is no less self-contradictory than (1), since M cannot be nonsuccessively past, present, and future.

The response of the tenser to this argument is to question McTaggart's original assumption, namely, that the tensed theory of time implies that (1) is true of each event E. This response seems warranted, since McTaggart provides no justification for this assumption. He simply asserts it. As I have stated, McTaggart admits that "the attribution of the characteristics past, present and future to the terms of any series leads to a contradiction, *unless it is specified that they have them successively*" and goes on to claim that the tensed theory of time is incoherent on the grounds that the contradiction expressed in (1) cannot be eliminated by (2), since (2) entails the contradiction (3); due to the reappearance of the contradiction, "the first set of terms [the events] never escape from contradiction at all."[37] But this claim is warranted *only if it is assumed* that the tensed theory of time entails that the terms possess the three properties nonsuccessively. Without this initial assumption, there is no contradiction in the first set of terms that needs to be escaped. The tenser, of course, rejects this initial assumption; according to him, the tensed theory of time entails not (1) but

> (4) E will be past, is now present, and was future; or E is now past, was present, and was (still earlier) future; or E is now future, will be present, and will (still later) be past.

Item (4) is consistent, since the properties of pastness, presentness, and futurity are ascribed to E at different times. For example, the first disjunct says that E possesses pastness in the future, presentness in the present, and futurity in the past. This implies an infinite regress, but the regress is not the vicious infinite regress that McTaggart assumes to obtain. Consider, for example, the clause "E is now present." According to my analysis, this asserts that the property of presentness inheres in the event E and also in its own inherence in E. This analysis may be understood as a way of responding to the question, "When does the property of presentness inhere in E?" Manifestly, if E is present, this property does not inhere in E in the past or the future but in the present. This means that presentness, not pastness or futurity, inheres in the inherence of presentness in E. A similar question can be raised about the latter inherence. When does presentness inhere in its own inherence in E? The answer is *at present*. This means that presentness not only inheres in E and in its inherence in E but also in its inherence in its inherence in E. This regress continues infinitely but in a benign manner, since at no stage in the regress is there a contradiction. This regress is also compatible with the other regresses that obtain if E is present. We need only consider one of them, that implied by "E will be past." If E *is now* present, then it *will be* past. This latter clause means that futurity inheres in the pastness of E. But when does futurity inhere in the pastness of E? The answer is already implicit in the present tense of "inheres" in the statement that "futurity inheres in the pastness of E." Futurity *now* inheres in the pastness of E. "E will be past" implies that "E *is now* such that it will be past." In terms of property inherences, this means that presentness inheres in the inherence of futurity in the inherence of pastness in E. But when does presentness inhere in the inherence of futurity in the inherence of pastness in E? *At present,* since it is *right now* that the pastness of E is now future. The rest of this regress, it is apparent, is a regress of the inherences of presentness in its own inherences. This bespeaks the predominance of presentness in the regresses, which is to be expected, since if anything is past or future it is *presently* past or future; and if anything is present, it is present *at present*. Despite the complexity of these regresses, I think it is intuitively clear that the two regresses described are compatible with each other and can obtain simultaneously:

1. *First Regress* (implied by "E is now present"). Presentness inheres in E, and in its own inherence in E, and in its own inherence in its inherence in E, and so on infinitely.
2. *Second Regress* (implied by "E will be past"). Pastness will inhere in E, and futurity inheres in the inherence of pastness in E, and presentness inheres in the inherence of futurity in the inherence of pastness in E, and presentness inheres in its own inherence in the inherence of futurity in the inherence of pastness in E, and presentness inheres in its own inherence in its own inherence in the inherence of futurity in the inherence of pastness in E, and so on infinitely.

The key idea here is that the property of presentness *inheres in the inherences of A-properties.* This idea has gone unnoticed in other discussions of McTaggart's paradox; and for this reason, the "logic of A-properties" has not been satisfactory understood. This failure goes hand in hand with the broader failure to develop the "logic of reflexive properties in general," where a reflexive property is one that inheres in its own inherence in something. Presentness is only one instance of a reflexive property, alongside such properties as identity, difference, oneness, being something, and the like. Consider, for example, the sentence

(5) The property of self-identity *inheres in* (is exemplified by) John.

Let us examine the property tie of inherence that ties self-identity to John. Surely, this property tie is *different from* the property of wisdom and is *identical with itself.* But if it is identical with itself, that entails

(6) The property of self-identity inheres *in its own inherence in* John.

Not only is John self-identical, but the inherence of self-identity in John is itself self-identical. But consider now the property tie of inherence that ties self-identity to its inherence in John and makes it true that the inherence of self-identity in John exemplifies self-identity. Surely, the property tie in question is identical with itself. This entails

(7) The property of self-identity inheres in its inherence in its inherence in John.

A benign infinite regress ensues. The statement of this infinite regress is, of course, not a part of the meaning of "John is self-identical," that is, a part of the proposition expressed by this sentence, since we do not grasp an infinite number of ascriptions of self-identity when we understand this sentence; but the statement of this regress is *entailed* by this sentence. John cannot be property-tied to self-identity if property ties are not identical with themselves.

Consider now

(8) The property of oneness inheres in John.

Surely, the property tie that ties oneness to John is *one* property tie, not two. In that case, (8) entails

(9) The property of oneness inheres *in its own inherence* in John.

It is easy to see that a benign infinite regress ensues. The same holds for other reflexive properties, such as the dyadic property of difference. (Surely, the inherence of difference in x is itself different from items other than the inherence of difference in x.) It seems to me that the key to understanding the regress implied by A-properties is just this notion of a reflexive property, for presentness is a reflexive property.

To substantiate my analysis and argument against McTaggart further, I will consider the views of Oaklander, Dummett, Mellor, and Schlesinger. I begin with Oaklander, since he has responded to an earlier statement of my criticism of McTaggart by claiming that my theory reintroduces McTaggart's contradictions at the level of property inherences. Oaklander writes:

> In order to avoid the difficulty of E's being simultaneously past, present and future, Smith is forced to claim that *the inherence of a temporal property in E is simultaneously past, present and future.* . . . . Thus, the notion of succession, analyzed in terms of the tensed inherence relations, does not really avoid the contradiction of something being past, present and future since it rearises at the level of inherence.[38]

Oaklander regards the contradictory and nonsuccessive possession of the three temporal properties to be a simultaneous, rather than a timeless, possession. This interpretation may be adopted for the sake of responding to Oaklander's argument.

Let me point out, to begin with, that this passage is misleading insofar as it suggests that I *claimed* that the inherence of a temporal property is simultaneously past, present, and future, since a perusal of the article Oaklander is criticizing will show that this claim is nowhere made. But this passage can be more charitably interpreted as meaning merely that my theory *implies* that the inherence is simultaneously past, present, and future even though I did not make this claim myself. But why does Oaklander think that my theory implies this? Precisely because he adopts the unjustified assumption originally made by McTaggart, that the tensed theory of time entails that pastness, presentness, and futurity are possessed simultaneously. Oaklander assumes that

(10) The inherence of presentness in E is now present

entails

(11) The inherence of presentness in E is simultaneously past, present, and future.

The fact that Oaklander assumes this is evinced by his assertion that "if inherence is present, then it must be past and future as well,"[39] which Oaklander regards as implying that "the first order inherence relation has incompatible temporal properties simultaneously."[40] *But Oaklander gives no justification for this assumption.* Like McTaggart, he simply *asserts* that whatever possesses the three temporal properties must possess them simultaneously. By importing this unjustified and foreign assumption into the tensed theory of time, Oaklander, like McTaggart, proceeds to deduce the incoherence of the tensed theory. But this assumption is not a part of the tenser's theory of time. According to this theory, something possesses the three incompatible properties only successively. If the inherence of presentness in E is present, that does not imply the self-contradictory sentence (11) but the self-consistent sentence

(12) The inherence of presentness in E is *now* present, *was* future, and *will be* past.

This entails an infinite regress; but as I have pointed out, this regress is benign; at no stage is there a contradiction.

Michael Dummett likewise begins with the question-begging assumption that A-properties are predicated nonsuccessively of any given event E. He commences by endorsing McTaggart's dictum that (as Dummett words it) "the predicates 'past', 'present', and 'future' involve a contradiction: for on the one hand, they are incompatible predicates, and on the other, to every event all three apply."[41] Dummett notes that the attempt to resolve the contradiction by advancing to a higher level, namely

$$\left\{ \begin{array}{l} \text{past} \\ \text{present} \\ \text{future} \end{array} \right\} \text{ in the } \left\{ \begin{array}{l} \text{past} \\ \text{present} \\ \text{future} \end{array} \right\}$$

is of no avail, since there are contradictions on this level as well, for example, an event cannot be both "present in the present" and "present in the future." The same applies if we advance to still higher levels, and so on ad infinitum. But Dummett makes no attempt to justify his assumption that the tensed theory of time entails the initial contradiction that each event is nonsuccessively present, past, and future.

D. H. Mellor begins with the assumption that the tensed theory of time entails two logically incompatible theses, namely,

(13) Pe & Ne & Fe

and

(14) Pe $\rightarrow$ ~Ne; Pe $\rightarrow$ ~Fe; Ne $\rightarrow$ ~Fe; Ne $\rightarrow$ ~Pe; Fe $\rightarrow$ ~Pe; Fe $\rightarrow$ ~Ne,

where P is pastness, N, nowness or presentness, and F, futurity and $\rightarrow$ means "entails" and ~ means "not"). As Mellor sees it, the attempt to reconcile (13) and (14) by specifying different past, present, or future times at which e possesses P, N, and F results in more complex temporal properties; and some of the latter are themselves incompatible. This specification results in the expansion of (13) into

(15) PPe & PNe & PFe & NPe & NNe & NFe & FPe & FNe & FFe.

But (15) is no less self-contradictory than (13), since some of the conjuncts of (15), such as PNe and FNe, are incompatible. The resort to even more complex properties, such as FPN and PFN, is of no avail, since many of these more complex properties are themselves incompatible. The regress of more and more complex properties is vicious, since at every level of complexity some of the properties are mutually incompatible.

But the tenser, of course, rejects Mellor's question-begging assumption that the tensed theory of time entails (13). There is no vicious regress, since there is no contradictory and nonsuccessive predication to begin with. The tensed theory of time cannot be refuted by assuming, without argument, that it entails a self-contradictory thesis that is not a part of this theory as formulated by its proponents and that its proponents deny is entailed.

But this is not entirely fair to Mellor; for, to his credit, he acknowledges that the tenser will regard the assumption of (13) to be question-begging:

> I must deal with the complaint that in symbolising McTaggart's argument I have myself begged the question against tenses. Specifically, in using "Pe" to say that e is past, I have left out the verb "is." By so doing I have tacitly treated the "is" in "e is past" as a tenseless copula, which is why e's being past, present and future appears to be contradictory. For in fact the verb "is" in "e is past" is tensed, i.e. it really means that e is *now* past. And given that, the contradiction vanishes, since if e is now past, it is not also now present or future. Of course, it *was* future and it *was* present, but that is quite compatible with e being now past.[42]

Mellor proceeds to argue that introducing a tensed copula cannot save the tensed theory of time from contradiction, since this introduction entails a vicious infinite regress of its own.

> Anyone who . . . says the "is" in "e is past" is present tense, so that "e is past" means "e is now past," will have to say what tense "is" then has in "e is now past." . . . If the "is" in "e is now past" is tensed, as in "e is past," the same vicious regress appears in the form of the verb itself. For "e is past," meaning "e is now past," must now also mean "e is now now past," in which again the "is" must be either tenseless or present tense. If . . . present tense, the regress continues with "e is now now now past," "e is now now now now past," and so on *ad infinitum*. . . . To stop and give a definite answer at any stage only produces a contradiction, because if the sentence is true (at some present time) it is also false (at some other). The only way to avoid contradiction is never to stop at all, which is tantamount to admitting that the original sentence type has *no* tensed truth conditions, i.e. cannot be made true or false by any tensed fact such as that e is past, e is now past, e is now now past, etc.[43]

Several comments are in order. First of all, even if the premises of this argument are true, the argument is invalid; for if "one never stops at all," that does not entail that the original sentence has *no* tensed truth conditions but merely that it has no *finite* tensed truth conditions. If one never stops, the entailment is that the sentence has an infinite tensed truth condition. The tenser, if he accepts Mellor's premises, may concede to Mellor that the sentence is not made true by any single tensed fact of the form "e is now past" or "e is now now past," since tensed facts of this form are one and all finitely complex. But this concession is consistent with holding either (a) that the sentence is made true by a single tensed fact of the form "e is past now now now . . . ," which is infinitely complex, or (b) that the sentence is made true by an infinite number of finitely complex tensed facts of the form "e is now, . . . , now past," where the number of nows in each such fact is less than aleph-zero. This infinite factual complexity avoids contradiction, since it involves a specification of when each temporal property is exemplified: e exemplifies pastness *now,* and e's exemplification of pastness exemplifies nowness *now,* and the exemplification of nowness by e's exemplification of pastness exemplifies nowness *now,* and so on without end. A contradiction is produced only if one stops somewhere in this regress. For example, if one stops after the exemplification of nowness by e's exemplification of pastness, it will not be specified when e's exemplification of pastness exemplifies nowness; one will thereby not be able to say that the exemplification occurs now, rather than in the past or future; and the inference that it tenselessly obtains in the past, present, and future cannot be blocked. But if one never stops, it is specified when this exemplification occurs and when each other exemplification of a temporal property occurs, that is, now, rather than in the past or future. Thereby, one avoids predicating incompatible properties of any one of the exemplifications.

But this response concedes too much to Mellor. First of all, his assumption is false that if "e is past" is tensed, it means "e is now past"; for the the copula in "e is past" is not an indexical, whereas "now" is an indexical (as was argued in chapter 4), and no nonindexical has the same meaning as an indexical. The regress can be stated in terms of nonindexical A-locutions, such as "the being past of e is present," "the being present of the being past of e is present," and so on. Second, his assumption that the original sentence "e is past" means the same as the sentences stating the further stages of the regress is also false, since it confuses meaning with entailment. "E is past" does not mean "The being present of the being past of e is present" but entails this. If sentences meant what they entailed, then "John is running" would mean "John, who is running, is neither the number one, nor the number two, nor the number three, and so on ad infinitum," since "John is running" is true only if John is identical within himself and different from each number. Third, if we are to follow Mellor in holding that "the truth conditions of a sentence S" refers to *whatever* must obtain if S is true, then every sentence has infinitely complex truth conditions; for "John (is) walking at noon" is true only if $1 + 1 = 2$, $2 + 1 = 3$, $3 + 1 = 4$, and so on ad infinitum, and only if John is identical with himself, the property tie tying John and identity is identical with itself, and so on ad infinitum. But if we are to maintain this inflated notion of a truth condition, we must be careful to distinguish between the state of affairs that corresponds to a sentence (as used on some occasion) and the states of affairs whose obtaining is the necessary condition of the truth of the sentence. "John (is) walking at noon" does not correspond to the infinite number of states of affairs in the series $1 + 1 = 2, 2 + 1 = 3$, . . . but to the single, finitely complex state of affairs consisting of John, walking, and noon (and arguably some property that is partly composed of presentness) as ordered by property ties. Likewise, "e is past" does correspond not to an infinite number of inherences of presentness in its own inherences but to presentness' inherence in the pastness of e. "E is past" expresses a finitely complex proposition whose constituents bear correspondence relations (identity, replacement, or instantiation) to a finitely complex state of affairs, although this finite state of affairs cannot obtain unless an infinite number of other states of affairs also obtain.

I wish to consider, finally, George Schlesinger's version of McTaggart's paradox. Schlesinger states the alleged initial contradiction in terms of instantaneous events. He allows that it is possible for a temporally extended particular E to have a property Q at one moment and to lack that property at the next moment, since there are at least two points of time occupied by E and E could possess the property Q at one of these moments and not at the other. But suppose that E is an instantaneous event. It cannot then acquire and lose any property Q: "Because E is not extended, that is, there are no two points in time occupied by it, at one of which it could have Q and at the other it could lack it, E simply does not have the scope to accommodate the incompatible properties Q and not Q."[44] It follows, therefore, that no instantaneous event E can acquire and then lose properties of futurity, presentness, or pastness.

But this McTaggart-based argument is another instance of a petitio principii. Schlesinger's argument has this structure:

(16) For any event E, E acquires or loses properties only while it occurs.

(17) Any event E can acquire or lose a property only if it occurs for at least two instants.

Therefore,

(18) Instantaneous events cannot acquire or lose any property.

Therefore

(19) Instantaneous events cannot acquire or lose A-properties.

Schlesinger, however, makes no attempt to justify (16); and (16) is just the thesis that the A-property theory is false, for the A-property theory is

(20) For any event E, E first has the property of futurity (when it is not yet occurring), then has the property of presentness (when it is occurring), and finally has the property of pastness (when it is no longer occurring).

Since Schlesinger assumes, rather than proves, (20) to be false, his argument is informally invalid.[45] Of course, Schlesinger is himself an A-theorist, and goes on to suggest a "resolution" of this alleged paradox by postulating a second time series in which all the moments of the first time series are present; but Oaklander has successfully argued that Schlesinger's "resolution" is in fact self-contradictory.[46] I would argue that a "resolution" was not needed in the first place, since the problem Schlesinger invented is spurious.

I conclude that considerations based on McTaggart's paradox do not impugn the A-property theory, since they are based on question-begging assumptions. That this is, indeed, a common failing of the McTaggart-based arguments is further suggested by J. M. Shorter's recent article "The Reality of Time," in which he argues for the falsity of the A-property theory on the basis of the un-argued-for assumption that the A-property theory entails the contradiction that "every event is (tenselessly) past, present and future."[47] This is no more convincing than Paul Horwich's more recent McTaggart-based argument, based on the same un-argued-for assumption that the A-property theory entails the contradiction that "every event and time has [nonsuccessively] the qualities of *past, present, and future.*"[48]

Notes

1. F. P. Ramsey, *The Foundations of Mathematics* (New York: Humanities, 1950), 116.

2. John Searle, *Speech Acts* (Cambridge: Cambridge University Press, 1969), 102–3.

3. Joseph Almog, "Naming Without Necessity," *Journal of Philosophy* 83(1986):238, n. 18.

4. Bertrand Russell, *Principles of Mathematics* (New York: Norton), 49–50.

5. Roderick Chisholm, *Person and Object* (London: George Allen & Unwin, 1976), 118.

6. P. F. Strawson, *Individuals* (New York: Anchor Books, 1959), 167.

7. Gilbert Ryle, "Plato's Parmenides," *Mind* 1939: 138; William Winslade, "Russell's Theory of Relations," in *Essays on Bertrand Russell*, ed. E. D. Klempke (Chicago: University of Illinois Press, 1970), 91; Richard Parker, "Bradley's Paradox and Russell's Theory of Relations," *Philosophy Research Archives* 10(1985): 262; Alan Donagon, "Universals and Metaphysical Realism," *Monist* 47(1962–63): 225.

8. Alexius Meinong, "On Objects of Higher Order and Their Relationship to Internal Perception," in *Alexius Meinong* ed M. S. Kalsi (The Hague: Martinus Nijhoff, 1978), 147; Russell, *Principles of Mathematics*, 50–52; Nicholas Wolterstorff, *On Universals* (Chicago: University of Chicago Press, 1970), 94ff.

9. Roderick Chisholm, "Events and Propositions," *Noûs* 4(1970): 15–24; idem "States of Affairs Again," *Noûs* 5(1971): 179; Nicholas Wolterstorff, "Can Ontology Do Without Events," in *Essays on the Philosophy of Roderick Chisholm*, ed. Ernest Sosa (Amsterdam: Rodopi, 1979), 177–201; Alvin Plantinga, *The Nature of Necessity* (Oxford: Clarendon, 1974).

10. Plantinga, *Nature of Necessity*, 44.

11. *Ibid.*, 45.

12. A. N. Prior, *Papers on Time and Tense* (Oxford: 1968); Ferrel Christensen, "The Source of the River of Time," *Ratio* 18(1976): 131–43; A. B. Levison, "Events and Time's Flow," *Mind* 96(1987): 341–53; G. Lloyd, "Time and Existence," *Philosophy* 53(1978): 215–28; W. L. Craig, *The Only Wise God* (Grand Rapids: Baker Book House, 1987).

13. Prior, *Papers on Time and Tense*, p. 4.

14. Christensen, "The Source of the River of Time," 137.

15. C.J.F. Williams, *What Is Existence?* (Oxford: Clarendon, 1981), 109–10.

16. Michael Dummett, *Frege*, 2d. ed. (Cambridge: Harvard University Press, 1981), p. 387.

17. Ibid.

18. Nicholas Wolterstorff, "Can Ontology Do Without Events?," in *Essays on the Philosophy of Roderick Chisholm*, ed. Ernest Sosa (Amsterdam: Rodopi, 1979), 190.

19. Ibid.

20. Ibid.

21. Idem, "McTaggart's Paradox and the Nature of Time," *Philosophical Quarterly* 24(1974): 297, 299.

22. L. Nathan Oaklander makes a similar point against Christensen in his penetrating criticism of Christensen's theory in *Temporal Relations and Temporal Becoming* (New York: University Press, America, 1984), 90–92.

23. Levison, "Events and Time's Flow," 352.

24. Lloyd, "Time and Existence," 228.

25. A. N. Prior, *Past, Present, and Future*, (Oxford: Clarendon, 1967), 18.

26. D. H. Mellor, *Real Time* (Cambridge: Cambridge University Press), 95.

27. Christensen, "McTaggart's Paradox," 297.

28. Frege writes, for example, that the sentence "There is at least one square root of 4" is equivalent in meaning to "The concept *square root* is realized" (*Translations from the Philosophical Writings of Gottleibe Frege* ed. Peter Gesch and Max Black (Oxford: Basil Blackwell, 1977) 50; and Russell writes: "When you take any propositional function and assert of it that it is possible, that it is sometimes true, that gives you the fundamental meaning of 'existence'. . . . Existence is essentially a property of a propositional function" (*Logic and Knowledge,* New York: G. P. Putnam's Sons, 1956, 232).

29. Ernest Sosa, "The Status of Becoming: What Is Happening Now"," *Journal of Philosophy* 77(1979), 28.

30. Prior, *Past, Present, and Future*, 14.

31. Christensen, "McTaggart's Paradox," 29.

32. J.M.E. McTaggart, "Time," in *The Philosophy of Time*, ed. Richard Gale (New York: Anchor Books, 1967), 96.

33. Pavel Tichy, "The Transiency of Truth," *Theoria* 46(1980): 26–42.

34. L. Nathan Oaklander, "McTaggart's Paradox and the Infinite Regress of Temporal Attributions: A Reply to Smith," *Southern Journal of Philosophy* 25(1987): 425–31.

35. See nn. 30–31 and C. D. Broad, *An Examination of McTaggart's Philosophy*, vol. 2 (Cambridge: Cambridge University Press, 1938), part 1.

36. See n. 34.

37. See n. 32.

38. L. Nathan Oaklander, "McTaggart's Paradox," 430.

39. Ibid., 429.

40. Ibid., 430.

41. Michael Dummett, *Truth and Other Enigmas* (Cambridge: Harvard University Press, 1978), 351.

42. Mellor, *Real Time*, 96–97.

43. Ibid., 97–98.

44. George Schlesinger, *Aspects of Time* (Indianapolis: Hackett, 1980), 53.

45. David Zeilicovici has shown that Schlesinger makes a related question-begging assumption, that is, that events are property-closed. See David Zeilicovici, "A (Dis)solution of McTaggart's Paradox," *Ratio* 28(1986): 175–95.

46. L. Nathan Oaklander, "McTaggart, Schlesinger, and the Two-Dimensional Time Hypothesis," *Philosophical Quarterly* 33(1983): 391–97.

47. J. M. Shorter, "The Reality of Time," *Philosophia* 14(1984): 326.

48. Paul Horwich, *Asymmetries in Time* (Cambridge: Massachusetts Institute of Technology Press, 1987), 24.

6

Presentness as a Logical Subject of Tenseless Sentences

6.1 All Subjects but Presentness Are Local Logicometaphysical Subjects

In chapter 5 I argued that presentness is a logical subject of all A-sentences and a metaphysical subject of all A-states of affairs. This is perhaps the most initially plausible and easiest-to-grasp part of presentism, for the remaining doctrines do not have any significant degree of "intuitive obviousness" about them. Indeed, they seem, on the face of it, to be implausible. It seems counterintuitive to hold that all *tenseless* sentences, be they B-sentences, empirical generalizations, or analytic sentences, refer to presentness and ascribe to it some monadic property. It further seems that if there is any argument that shows presentness to be a logical subject of these sentences, similar arguments must show that other properties are also universal logical subjects, such as the trivially essential properties of self-identity, or oneness, or being something.

These two issues shall be the focus of this chapter. I will begin in the current section by arguing that nothing different from presentness is a universal logico-metaphysical subject. Specifically, I shall show that no concrete object, such as the universe or God (if he exists) is a universal subject, and that no trivially essential property is a universal subject. I shall discuss only these candidates, since these are the most likely candidates for universal subjects. If the universe and self-identity are not universal subjects, then surely the Empire State Building and the property of being blue are not universal subjects. In the remaining sections, I will show that presentness is a logical subject of all sentences other than A-sentences. Specifically, I shall argue that all so-called "tenseless sentences," such as B-sentences, sentences stating laws of nature, and mathematical sentences, refer to presentness and ascribe it some monadic property. I shall then argue (in chapter 7) that every possibly true sentence has absolute presentness for a logical subject, that is, ascribes to presentness properties of such forms *inheres in such-and-such*, rather than of such forms as *inheres in such-and-such relatively to a reference frame R*.

I begin in this section with the negative argument that nothing (besides presentness) is a universal logico-metaphysical subject. The argument that all subjects but presentness are local subjects depends on the notion of a state of affairs. There are two basic ways to conceive states of affairs—as thick or thin—and the conception relevant to presentism is clearly the thin conception, since the thick conception entails that there are an infinite number of universal metaphysical subjects. But the thick conception has several species. One thick conception involves the idea that any state of affairs that includes a substance (person, star, table, etc.) as a part includes the substance *as exemplifying all its essential nonrelational properties*. Other thick conceptions represent the state of affairs as including the substance with all its *essential properties* be they relational or nonrelational, or all its *essential and nonessential nonrelational properties, or all its essential and nonessential properties*. These states of affairs include thick substances, that is, "substances as exemplifying all their essential nonrelational properties (or all their essential properties or all their nonrelational properties, etc.) Thick states of affairs also include thick abstract objects, such as properties. The property of running, conceived on one of the thick conceptions, is running *as exemplifying all its essential nonrelational properties*, such as being a property, being one, being a type of movement, and so on. "Thicker" conceptions result when we introduce essential or accidental relational properties, such as being different from the number one or being different from John.

This enables us to see that thick states of affairs have many universal metaphysical subjects. Among the many universal metaphysical subjects of thick states of affairs are oneness, being a property, and being something, as well (if we use the thick conception that includes all essential n-adic properties) as being self-identical, being different from the number two, being different from the property of wisdom, and the like. The reason each of these properties is a universal metaphysical subject of thick states of affairs is that every state of affairs includes at least one property as a part, and each property essentially possesses monadic properties of being a property, being one, being something, and so on. These are metaphysical subjects, since (1) they are parts of these thick states of affairs, (2) they have the role in these states of affairs of being property possessors, and (3) they belong to some thick referent of any sentence-token that corresponds to the state of affairs. For example, the property of oneness is a metaphysical subject of the thick state of affairs consisting of John as running; for it belongs to this state of affairs as directly possessing the property of *inhering in running* and as directly possessing the property of *inhering in John*. Oneness is a part of thick John and a part of thick running and thus is a part of the thick referents of any token of "John is running" or a synonymous sentence that corresponds to the state of affairs. Note, further, that if we consider thick states of affairs as including substances and abstract objects as possessing all their properties, then every concrete object (substance or first-order event) and every abstract object is a metaphysical subject of every state of affairs, for every object O possesses the property of *being different from x,* for every other object x, so that each object x belongs to the state of affairs as possessing *being different from O*. Indeed, on this view, there would seem to be at each present time only one state of affairs S, such that every concrete and abstract object was a metaphysical subject of S; every property tie that in fact obtained was an ordering tie in S; and every proposition that was true corresponded to S.

It is obvious, then, that presentism is a theory about thin states of affairs. Such states of affairs include thin concrete objects and thin abstract objects. The thin state of affairs corresponding to a true token of "John is running" includes thin John and thin running, as well as thin presentness. Thin John is John taken apart from all his properties, John taken solely as the particular that exemplifies John's properties, and without these properties themselves. This does not, of course, mean that thin John is a "bare particular" in the sense of a particular that can exist by itself without any properties. It means, rather, that (a) this particular exists in a thin state of affairs S, such that no properties of this particular belong to S except thin running and (b) the thin state of affairs containing this thin particular cannot obtain unless many other thin states of affairs also obtain, namely, all those that include thin John as exemplifying his essential properties. The thin state of affairs consisting of thin John as presently exemplifying running cannot obtain unless there obtains another state of affairs containing thin John as exemplifying humanity, and so on for all the other essential properties of John. This does not imply that there are many numerically distinct thin Johns, one for each different thin state of affairs that contains thin John; rather, there is only one thin John, and he exists in many different thin states of affairs. The thin John that is a part of John-as-running is numerically identical with the thin John that is a part of John-as-human. These two thin states of affairs are partially identical in that they both include thin John, but they are also partially different in that one includes thin running but not thin humanity and the other includes thin humanity but not thin running.

Thin John may be considered as a continuant or substance, such that numerically one and the same particular, thin John, continues (persists) from one moment to the next. If we adopt this conception, then we shall be working with a *substance* ontology, rather than a *temporal parts* ontology. On a temporal parts ontology, there is no thin particular (thin John) that continues from one moment to the next but, rather, a temporal series of thin particulars, each of which exists at one moment only; and "thin John" will, in this case, refer to the series of these particulars. Each of these thin particulars will be a temporal part of thin John. In this case, we should not say that thin John is a part of any momentary state of affairs but, rather, that a temporal part of thin John is a part of a momentary state of affairs. In this work, I am assuming a substance ontology. This is in keeping with the traditional alignments, since proponents of the tensed theory of time (e.g., Prior and Chisholm)[1] usually assume a substance ontology and proponents of the tenseless theory of time (e.g., Smart and David Lewis,[2] with the notable exception of Mellor)[3] usually assume a temporal parts ontology. (I would note, however, that I cannot accept the belief that there is a logical connection between the tensed theory and the substance theory or between the tenseless theory and the temporal parts theory, although it is not my purpose to argue this here.[4] I believe that presentism could be developed within the framework either of a substance ontology or a temporal parts ontology; but for the sake of definiteness, I am assuming only one of these ontologies in this work, namely, the substance ontology.)

I indicated that thin states of affairs include thin properties, as well as thin substances. The thin running that is a part of the thin state of affairs presentness-as-inhering-in-the-running-of-John is the property of running taken apart from its properties. Just as thin John exists as a part of this thin state of affairs only if there simultaneously obtain all the thin states of affairs containing thin John as exemplifying his essential properties, so thin running exists as part of this thin state of affairs only if there simultaneously obtain all the thin states of affairs, including thin running as exemplifying all its essential properties (e.g., being a property, being self-identical, being a type of movement, being different from the property of walking). Likewise, thin presentness belongs to this thin state of affairs of presentness-as-inhering-in-John's-walking only if there simultaneously obtain all the thin states of affairs consisting of presentness as possessing its essential properties.

The thesis of presentism is that thin presentness is the only thin object that is a universal metaphysical subject of every thin state of affairs. Presentism is a very fine grained theory about reality, in that it deals with thin, rather than thick, objects and states of affairs and involves the very fine grained distinctions that are required for a theory about thin phenomena. But asserting these fine-grained theses is one thing, establishing them, another. In the remainder of this section, I shall show that no thin object different from presentness is a universal metaphysical subject of thin states of affairs. Since presentism is a theory about reality thinly conceived, I shall henceforth use "state of affairs" and "metaphysical subject" as synonymous with "thin state of affairs" and "metaphysical subject of a thin state of affairs," respectively.

The crucial negative part of presentism is to show that God (if he exists), the concrete world whole, and trivially essential properties are not universal metaphysical subjects. It is obvious that no ordinary concrete object—no star, animal, or person—is a subject of every state of affairs, so I shall not bother to argue this. (One to whom this is not obvious is thinking of thick states of affairs, rather than the thin ones I have in mind.) It is less obvious that God, if he exists, is not a universal metaphysical subject; and it is less obvious that the concrete world whole, the aggregate of all existents ("the universe," if your prefer),[5] is not a universal metaphysical subject. I shall argue shortly that they are not universal subjects. Regarding abstract objects, it is obvious that properties such as running or blueness are not universal metaphysical subjects; but it is less obvious that trivially essential properties (e.g., self-identity and oneness) are not universal subjects, and my subsequent arguments will concern them alone.

My argument that God (if he exists), the concrete world whole, and trivially essential properties are local metaphysical subjects is most easily made via an argument that they are local logical subjects, for something is a universal metaphysical subject only if it is a universal logical subject (see sections 5.1, 5.6). It belongs to the definition of a metaphysical subject that it is a part of a state of affairs—a property possessor in that state of affairs—and that it is a logical subject of any sentence-token that corresponds to that state of affairs. Consequently, if it can be shown that God or self-identity is not a logical subject of some sentence-token that corresponds to some state of affairs, it will follow that God or self-identity is not a metaphysical subject of that state of affairs and, eo ipso, not a universal metaphysical subject.

I shall argue that God, the concrete world whole, and trivially essential properties are not logical subjects of true tokens of "John is running" and therefore are not universal metaphysical subjects. We know from our earlier discussions that something is a logical subject of a sentence-token if and only if it is referred to by a part of that sentence-token and is ascribed a monadic property by a part of that token. But this characterization of a logical subject is too vague to enable us to decide whether or not God or trivially essential properties are universal logical subjects. The notion of a logical subject can be made more precise in terms of a definition of reference, for something is a logical subject only if it is a referent of a sentence-token. In keeping with the thin conceptions pertinent to presentism, I use "logical subject" to express a *thin* conception and, accordingly, require a corresponding definition of *thin reference*. We begin with the partial definition

> (1) L is a logical subject of a sentence-token S only if L is, or is a part of, the sole thin referent of a distinct syntactical part of S.

A thin referent is best discussed in terms of the more familiar notion of a thick referent. (When philosophers talk of the "referent" of a locution, they usually mean the thick referent in one of the four senses of thickness just distinguished). According to one conception of thickness, the thick referent of a token of "John" is *thick John* in the sense of thin-John-as-exemplifying-all-his-essential-nonrelational-properties; and the thick referent of a token of "running" is *thick running* in the sense of thin-running-as-exemplifying-all-its-essential-nonrelational-properties. Consider now the *nucleus* of a thick referent:

> (2) N is the nucleus of a thick referent if and only if N is both a part of the thick referent and exemplifies all the other parts of the thick referent.

The parts of the thick referent of "John" are thin John, oneness, humanity, walking-or-not-walking, and so on. The nucleus is thin John, since thin John is that which exemplifies oneness, humanity, walking-or-not-walking, and all the other essential nonrelational properties of John. The nucleus of the thick referent of "running" is thin running, for this exemplifies all the essential nonrelational properties of running. We may say, accordingly, that the thin referent of a part of a sentence-token is the nucleus of its thick referent. The thick referent of "John" is thin-John-as-exemplifying-all-his-essential-nonrelational-properties, and the thin referent of "John" is the nucleus of this thick referent, namely, thin John.

Given the partial definition (1) of a logical subject, we may give a sufficiently precise formulation to the question whether or not God, the concrete world whole, or trivially essential properties are logical subjects of a true token of "John is running." We may begin with the question, Is God or the concrete world whole either the nucleus of the thick referent of "John" or a part of this nucleus? The answer is negative. This nucleus is thin John, and God and the world whole are neither identical with thin John nor parts of thin John. Is God or the world whole either the thin referent of "running" or a part of this thin referent? Obviously not. The thin referent is thin running; and neither God nor the world whole are identical with, or parts of, thin running. Is God or the world whole either a thin referent (or a part thereof) of the present tense of the copula or the pure copula? Obviously not. The pure copula has no referent, since its semantic contribution is solely to *convey* the propositional relatedness of John to running. The thin referent of the present tense of the copula is presentness; and God and the world whole are neither identical with, nor parts of, thin presentness. It follows, then, that God and the world whole are not logical subjects of a true token of "John is running" and, eo ipso are not metaphysical subjects of the corresponding state of affairs. This, in turn, entails that they are not universal metaphysical subjects.

Are trivially essential properties, such as oneness and self-identity, logical subjects of tokens of "John is running"? Clearly, oneness and self-identity are parts of the thick referents of "John," "running," and the present tense of the copula, since they are among the essential nonrelational properties of John, running, and presentness. (If self-identity is considered a relational property, then it would be excluded.) But they are neither identical with, nor parts of, the thin referents of these tokens. True, they are *properties of* these thin referents; but that does not entail they are *parts of* these thin referents. "F is an essential nonrelational property of the thin referent of 'T'" entails that "F is a part of the thick referent of 'T'," but it does not entail that it is a part of the thin referent.

It may be granted that "Thin running exemplifies self-identity" does not entail "Thin running is partly composed of the property of self-identity"; but it may be argued that there is another reason for thinking self-identity is part of thin running, namely, that self-identity is a part of every thin property. It may be argued that every thin property is complex conjunction of other properties and includes as one of its conjuncts the property of self-identity. Thus, thin running is the conjunction of such properties as being in motion, having legs, being in time, being in space, being animate, being a substance, being one, being self-identical, being red or nonred, and the like. The proof of this, it might be said, is that nothing can possess thin running unless it possesses each of these other properties.

I would respond to this argument by pointing out that it confuses "is a property entailed by the property F" with "is a part of the property F," where one property F entails another property G if it is logically impossible for anything to possess F and not possess G. If each property is partly composed of each property it entails, then each property is infinitely complex; for each property entails an infinite number of other properties. For example, nothing can possess running unless it possesses being different from the number one, being different from the number two, and so on, ad infinitum. But the consequence that each property is infinitely complex is incompatible with the fact that we grasp, or are acquainted with, many properties and many propositions of which properties are parts; for infinities are not objects of acquaintance. We are not acquainted with infinitely complex properties; but we can refer to them via finitely complex properties with which we are acquainted and that are expressed, in part, by universal quantifiers. For example, we can refer to an infinitely complex property by means of grasping the finitely complex property expressed by "the property whose parts are all the properties of the form *being different than N,* for each number N." This "principle of finitude" is one that Castañeda has often emphasized. He notes:

> We believe and know many propositions to be true, but we do not believe, much less know, all the infinitely many consequences of what we know, or believe. . . .
> The fundamental law is that no belief and no episode of awareness has an infinite

content, i.e. an infinite proposition. . . . The infinities we deal with are never presented as such to consciousness; *infinities are thinkable, not in person, but through means, themselves finite, like quantifiers, properties and schemata, which point to them.*[6]

The conclusion I draw from this "principle of finitude" is that the propositions with which we are acquainted contain a finite number of parts and that the properties that partly compose these propositions are not conjunctions of all the properties (e.g., the trivially essential properties) that they entail. This conclusion is consistent with the thesis that what we grasp in all cases are *parts* of infinitely complex conjunctive properties and propositions, for finitely complex conjuncts of infinite conjunctive properties and propositions are themselves properties and propositions. If *being F and G and H and I and so on ad infinitum* is a property, then *being F* is a property, *being G* is a property, and so on. The conclusion I am drawing is merely that the properties expressed by the relevant parts of our sentence-tokens are finitely complex properties, even though they may be conjuncts of infinitely complex properties that are not expressed by parts of our sentence-tokens. Analogously, I would grant that what is expressed by a token of "John is running" is a conjunct of some infinitely complex proposition but would insist merely that this conjunct is itself a proposition, a finitely complex one. If *that p_1 and that p_2 and that p_3 and so on ad infinitum* is a proposition, then so is *that p_1.*

I believe that these considerations suffice to show that self-identity, oneness, and other trivially essential properties are not parts of each property. If trivially essential properties are part of each property, the reason for this would be that they are entailed by each property; but this is not a valid reason, since this same reason would warrant the conclusion that all the entailments of a property are parts of it, and this conclusion implies the false thesis that all properties have an infinite number of parts.

I conclude, accordingly, that although trivially essential properties are *exemplified* by the thin referents of "John," "is," and "running" and are *entailed* by the thin referents of "is" and "running," they are not *parts* of these thin referents and eo ipso are not logical subjects of tokens of "John is running." It follows that trivially essential properties are not metaphysical subjects of states of affairs that correspond to tokens of this sentence and therefore that trivially essential properties are not universal metaphysical subjects. Since there are no other plausible candidates for universal metaphysical subjects, the conclusion is suggested that presentness, if a universal metaphysical subject, is the only one.

It may be wondered whether property ties, namely, inherence and exemplification, are universal metaphysical subjects. After all, they belong to every state of affairs as the order among the parts of that state of affairs. But this already gives us the answer, for the first of the three defining characteristics of a metaphysical subject is that it is a *part* of a state of affairs (the other two being that it is a property possessor and a logical subject of any corresponding sentence-token). A part of a state of affairs is different from the *order* among the parts, and the property ties are orders of the parts. A property tie is a constituent, but not a part, of a state of affairs, just as a propositional relation is a constituent, but not a part, of a proposition. The property tie of inherence is not a part of presentness-as-*inhering-in*-John's running but is the constituent that orders presentness to another part, John's running. (On the difference between parts and constituents, see pp. 141, 151–158.)

Someone may object at this point that if my criterion (1) for being a logical subject rules out God, the concrete world whole, and trivially essential properties as universal logical subjects, then it also rules out presentness; for according to criterion (1), presentness is not a logical subject of past and future tensed A-sentence-tokens. Manifestly, presentness is not the sole thin referent of the past tense of the copula in "John was running." My response to this objection is to concede that presentness is not the sole thin referent of past and future tenses but to point out that it is a *part* of the thin referent of these tenses and that being a part of the sole thin referent suffices to make presentness a logical subject according to (1). The thin referent of the past tense is *present pastness* (thin presentness as inhering in the inherence in [something] of thin pastness), and this thin property includes thin presentness as a part.

But this conclusion might motivate the objection that my criterion for being a logical subject is not too strict but too latitudinarian, for it seems counterintuitive to say that presentness is a logical subject of "John was running." Indeed, it seems counterintuitive to say even that presentness is a logical subject of "John is running." The intuitive view is that John and only John is a logical subject of both these sentences. The intuitively correct definition of a logical subject seems to be

> D1 L is a logical subject of a sentence-token S if and only if L is the thick referent of a proper name or definite description that occupies the position of grammatical subject in S.

I would have no argument with somebody who adopted the definition D1. Definition D1 would require a change not in the theory I am espousing but merely in its terminological expression. Instead of saying that presentness is the universal logical subject, I would say that presentness is the universal X, where an X is whatever satisfies the definition I have associated with the phrase "logical subject" in this treatise. But I do not believe that it increases clarity to use "X" or invent some neologism to express this concept, since the phrase "logical subject" in its established philosophical uses (only one of which is captured in D1) has a semantic content or contents sufficiently similar to my concept. "Logical subject," in many established philosophical uses, expresses the concept of being a referent of a part of a sentence and of being ascribed a property by that sentence. Thus, it is appropriately to regiment this phrase to express the exact conception of a referent I have in mind. Indeed, this is how philosophers normally treat technical terms like "logical subject"; they adopt some commonly used technical locution close in meaning to the concept they have in mind and stipulate that in their writings this locution will express this concept. Given this, my use of "logical subject" to express my definition is appropriate (whereas it would be inappropriate to use, e.g., "illocutionary force" or "epistemic operator" to express this definition, since there is no commonality with these locutions).

Given the technical sense I have assigned to "logical subject," the question remains whether presentness *is* a universal logical subject; for it seems far from obvious that presentness is, or is a part of, the thin referent of a distinct syntactical part of *every* sentence. In the past several chapters I have concentrated solely on A-sentences. But what about tenseless sentences? Surely, they do not contain a tensed copula that refers to presentness or a property that includes presentness. But initial appearances may be deceptive.

6.2 Tenseless Singular and Existential Sentences Other Than B-Sentences

An example of a tenseless singular sentence that is not a B-sentence is

(1) Socrates is wise

and an example of a tenseless existential sentence other than a B-sentence is

(2) There are dinosaurs.

Sentence (1) must be distinguished from the present tensed A-sentence that has the same grammatical appearance; for the A-sentence is now false, considering that it is now true that Socrates was wise while he was alive but is not wise now, considering that he is not now living, conscious, and knowing things. Sentence (2) likewise, must be distinguished from the present tensed sentence that has the same grammatical appearance as (2); for the present tensed sentence, unlike (2), is now false.

Given that (1) and (2) are tenseless, it might be thought self-evident that they do not ascribe presentness or a complex property to which presentness belongs. Indeed, does not "S is tenseless" entail "S ascribes neither presentness nor any property to which presentness belongs"? This depends upon what it means to say a sentence is "tenseless." I concede that (1) and (2) are tenseless in some sense, specifically, in the sense that their *copula lacks a single tense*. A copula (or verb, if the sentence has no copula) is singly tensed if it has only one of the three tenses—past, present, or future. The copula in "John was running" is singly tensed, since it is past-tensed and neither future- nor present-tensed. (I include the past perfect tense as a species of past tense, and the future perfect tense, of future tense) Now the copulae in (1) and (2) are not singly tensed and in this sense are "tenseless"; but that is consistent with saying that they are "tensed" in some other sense, a technical sense that I shall define. Although "tense" is normally used to refer only to what I call single tenses—the past, present, and future tenses—I am extending this word to refer also to (what I shall call) *multiple tenses*. There are two basic species of multiple tenses, *disjunctive tenses* and *conjunctive tenses*. If the copula in a sentence of the form "x is F" is disjunctively tensed, then this sentence is synonymous with a sentence that includes a disjunction of two or three of the three singly tensed copula, for example, a sentence of the form "x was, is, or will be F." If the copula in a sentence of the form "x is F" is conjunctively tensed, then this sentence is synonymous with a sentence that includes a conjunction of two or three of the singly tensed copulae, such as a sentence of the form "x was, is, and will be F." The theory of conjunctively tensed copulae will be developed and stated with more precision in the following section. In the present section, I shall concentrate on disjunctively tensed copulae; for sentences (1) and (2), I shall argue, have disjunctively tensed copulae.

Since the notion of multiple tenses is unfamiliar, it is necessary first to explain this idea in terms of the ordinary notion of tense. The word "tense," as normally used, is not a purely syntactical notion but combines syntactical and semantic notions. *Websters' New World Dictionary* expresses this normal use when it characterizes "tense" as "any of the forms of a verb that show the time of its action or state of being."[7] "Form of a verb" is a syntactical notion, but "show the time of its action or state of being" is a semantic notion. We can, however, by a process of abstraction, form a purely syntactic notion of tense, namely, that it is *a form of a verb*. ("Verb" is being used here in a broad sense to include copulae.) Consider the "is" that occurs in the present tensed A-sentence "John is running" and the "is" that occurs in the so-called "tenseless" sentences "Socrates is wise," "The Battle of Waterloo is later than the Battle of Hastings," and "Triangularity is a shape." These four sentences contain a common syntactical element, the "is." Something syntactically common to the "is" in the four sentences is that it is in the *purely grammatical present tense*. The differences in the tensed or "tenseless" character of the "is" in the several sentences appear only when the semantic properties of the purely grammatical present tense are taken into account. The tense of the "is" in "John is running" is "the present tense" in the normal (syntactico-semantic) sense that it is the purely grammatical present tense qua possessing the semantic property of ascribing the A-property of presentness. The "is" in the other three sentences is "tenseless" in the normal (syntactico-semantic) sense that it is in the purely grammatical present tense and does not have the same semantic content as any one of the three tenses—past, present, or future. But this description of "tenselessness" provides a purely negative characterization of the semantic content of the "is" in the "tenseless" sentences. My argument is that the purely grammatical present tense in the "tenseless" sentences also has a certain positive semantic property, namely, of *ascribing some property of which presentness is a part,* such that this property is not identical with any one of the three A-properties. I shall argue, regarding "Socrates is wise," that the purely grammatical present tense has the positive semantic property of ascribing the disjunctive property *being past or present or future,* which, although it includes disjunctively all three A-properties, is not identical with any one of these A-properties.

(Some philosophers, such as D. M. Armstrong, reject the idea that there are disjunctive properties.[8] I do not find their arguments convincing; but in deference to their position, I will show how my talk of disjunctive properties can be paraphrased away. The statement " 'Socrates was, is, or will be wise' ascribes the disjunctive property *being past or present or future*" may be interpreted as meaning that this sentence about Socrates expresses a disjunctive proposition that includes the three A-properties but no disjunctive property. On this interpretation, "Socrates was, is, or will be wise" does not express the proposition *that Socrates exemplifies the property of being past or present or future* but, instead, expresses the proposition *that Socrates either exemplifies the A-property of pastness or exemplifies the A-property of presentness or exemplifies the A-property of futurity*. The latter proposition is a disjunction of three A-propositions and is true if one of these disjuncts is true. Presentism is consistent with either interpretation, since, in both cases, presentness is, or is a part of, a thin property ascribed by this sentence. But I shall henceforth talk of "the disjunctive property of being past or present or future" and shall leave it to the readers who side with Armstrong to paraphrase this talk in the manner suggested.)

Before I present my argument that the copulae in "Socrates is wise" and "There are dinosaurs" are disjunctively tensed, I should relate the remarks just made to the characterization of a logical subject offered in the previous section. I said in section 6.1 that something is a logical subject of a sentence (in my special technical sense) only if it is the thin referent—or a part of the thin referent—of a distinct syntactical element in the sentence. The thesis about presentness may be formulated in part as the thesis that every English sentence includes as a distinct syntactical element a purely grammatical present, past, or future tense and that each grammatical tense has the semantic feature of thinly referring to presentness or to a property of which presentness is a part. Accordingly, the thesis that the so-called tenseless sentences have presentness for a logical subject is tantamount to the thesis that they have a tense in a purely grammatical sense and that this tense thinly refers to a property partly composed of presentness.

My definition of the disjunctive tense is that it is a purely grammatical present tense that has the semantic property of ascribing *being past or present or future* or a disjunction of any two of these properties. But this definition is not sufficient to tell us whether any given sentence is disjunctively tensed. We need some criterion whose satisfaction shows that a sentence is disjunctively tensed. The following criterion C1 follows from my definition of the disjunctive tense:

> C1 A sentence S contains a disjunctively tensed copula if and only if S's copula is in the purely grammatical present tense and S is synonymous with a sentence S_1 that is identical with S except for the fact that S_1 replaces S's copula with a disjunction of the past, present, and future tensed copulae.

I believe the so-called tenseless sentence "Socrates is wise" has a disjunctively tensed copula, since this sentence is synonymous with "Socrates was, is, or will be wise." It is obvious, to begin with, that "Socrates is wise" *entails* that he was, is, or will be wise, since if it did not entail this, it would be possible for it to be true that "Socrates is wise and Socrates is never wise," which is an analytic contradiction (in Jerrold Katz's sense).[9] (I shall show in a later section that the alleged possibility of Socrates being wise "in eternity" or a "in a timeless world" does not defeat this entailment.) But there is also a synonymy relation between "Socrates is wise" and "Socrates was, is, or will be wise." Two sentences or sentence-tokens are synonymous if and only if they are logically identical, which should be understood in the thin sense discussed in section 6.1; S and S_1 are logically identical if and only if they thinly refer to all and only the same items and ascribe to these items the same thin properties. Now, "Socrates" thinly refers to Socrates in both sentences, "wise" thinly refers to wisdom, and their pure copula conveys that thin Socrates is propositionally E-related to thin wisdom (i.e., that thin Socrates exemplifies thin wisdom if the expressed proposition is true). The only syntactical difference between the two sentences is the grammatical tense of their copulae. To establish that the two sentences are synonymous, all that is needed is an argument that the purely grammatical present tense in the one sentence has the same thin semantic content as the disjunction of the purely grammatical past, present, and future tenses in the other sentence. I believe there is an argument for this, part of which is that the two sentences are intersubstitutable *salva veritate* in belief contexts. If it is true that

(3) John believes that Socrates is wise,

then it is also true that

(4) John believes that Socrates was, is, or will be wise.

It can never be the case that an utterance of (3) has a different truth value than an utterance of (4) unless these sentences are given some artificial meaning that they do not normally have.

Intersubstitutivity in belief contexts is a necessary but insufficient condition for synonymy (logical identity); for "Socrates, who is human, is wise" is intersubstitutable *salva veritate* with "Socrates is wise" in (3) and yet is not synonymous with it. (I am assuming here the use of "Socrates" as a name of the Greek philosopher and not, say, as a name of somebody's dog.) They are not logically identical, since "Socrates is wise" contains no distinct syntactical part that thinly refers either to the property of being human or to something that is partly composed of the property of being human. (The property of being wise is not partly composed of being human, since, if it were, "God is wise" would be analytically false.) To be sufficient for logical identity, intersubstitutivity in belief contexts must be supplemented with the condition of *syntactico-semantic correlation.* Two sentences or sentence-tokens S_1 and S_2 whose logical identity is in question or not yet known are syntactico-semantically correlated if and only if, *consistently with their known semantic properties,* there is some way of dividing S_1 and S_2 into syntactical parts, such that for each part P_1 distinguished in S_1 there is one and only one part P_2 in S_2 that has the same thin semantic content as P_1. The parts into which they are divided may be simple or complex, such that it is permissible to correlate a simple syntactical part of S_1 with a complex syntactical part of S_2 or vice versa. There is one way of dividing up the parts of "Socrates was, is, or will be wise" and "Socrates is wise" that produces a correlation. The former sentence can be divided into three parts, "Socrates," "was, is, or will be" and "wise"; and the latter sentence into the three parts "Socrates," "is," and "wise." The correlation is that "Socrates" has the same thin semantic content as "Socrates," "wise," the same content as "wise," and "was, is, or will be," the same content as "is." This correlation is consistent with the known semantic properties of these two sentences; for it is consistent with their intersubstitutivity in belief contexts, their logical equivalence, and the thin semantic contents of the expressions "Socrates," "wise," and "was, is, or will be." The only thing that is not antecedently known (and that is currently in question) is the thin semantic content of the purely grammatical present tense of the "is" in "Socrates is wise," and the assignment to it of the same thin semantic content that is possessed by the complex tense of "was, is, or will be" is consistent with all the aforementioned known semantic properties of the sentences.

A similar argument shows that the "tenseless" existential sentence "There are dinosaurs" is disjunctively tensed. If John believes that there are dinosaurs, then he believes that there were, are, or will be dinosaurs. Further, "There are dinosaurs" is syntactico-semantically correlated with "There were, are, or will be dinosaurs," since it is consistent with our antecedent knowledge of these two sentences that the purely grammatical present tense of "are" in the first sentence has the same thin semantic content as the purely grammatical tenses of "were, are, or will be." If we take the existential quantifier "There are" as expressing a second-order property, then "There are dinosaurs" means "*Being a dinosaur* is exemplified," which in turn is synonymous with "*Being a dinosaur* was, is, or will be exemplified." Thus, these sentences are synonymous with presentist sentences, such as "Presentness either inhered, inheres, or will inhere in Socrates' being wise" and "Presentness inhered, inheres, or will inhere in the exemplification of *being a dinosaur*."

6.3 B-Sentences

A B-sentence was defined in section 1.2 as including these characteristics:

  i. It does not contain a tensed copula or verb or a temporal indexical unless, perhaps, within a subclause that is in quotation marks.
  ii. It is explicitly or implicitly about an event (i.e., a *concrete event*, the exemplification of an n-adic property or properties by a substance or substances).
  iii. It refers directly or indirectly to the event's B-position or ascribes to the event a polyadic B-property but does not ascribe to it an A-property.
  iv. It is logically contingent but is not a universal generalization.

An example of a B-sentence is "The Battle of Waterloo is later than the Battle of Hastings," which refers to two concrete events, the two battles, and ascribes to them the polyadic property of *later*. My use of "refers," "ascribes," and "is about" in this definition (as elsewhere) is not a use of these verbs as verbs of success. For example, by "refers," I do not mean "succeeds in referring" but "purports to refer." Thus, "Bush's assassination is later than Carter's assassination" counts as a B-sentence, since this sentence does not succeed either in referring to B-related events or in being about some events (since there are no events for it to be about); but it nonetheless purports to refer to, or be about, some events. Correlatively, my thesis that presentness is a logical subject of each possibly true sentence means that each possibly true sentence purports to refer to presentness and ascribe it some property.

Point i in the definition should be understood as meaning that B-sentences do not contain a singly tensed copula or verb and are "tenseless" in this sense. As indicated in the last section, a copula is singly tensed if it has only one of the three tenses, past, present, or future. But this single "tenselessness" is consistent with being multiply tensed, either disjunctively or conjunctively. I shall argue that B-copulae are conjunctively tensed.

The following criterion enables one to decide whether or not a copula is conjunctively tensed. This criterion applies to simple sentences, by which I mean a sentence that contains only one copula.

> C2 A simple sentence S contains a conjunctively tensed copula if and only if (1) S's copula is in the purely grammatical present tense; (2) S entails at least two simple sentences S_1 and S_2 that are identical with S except for the two facts that (a) the copulae in S_1 and S_2 are singly tensed, with the tense of S_1's copula being different from the tense of S_2's copula, and (b) the copula in S_1 or S_2, but not in S, may be modified by a temporal adverb, such as "always"; (3) S is synonymous with some sentence S' that is identical with S except that the copula of S' is the conjunction of the different singly tensed (and perhaps modified) copulae that belong to the entailed sentences S_1, S_2, mentioned in condition 2.

According to C2, the B-sentence

(1) The Battle of Waterloo is later than the Battle of Hastings

has a conjunctively tensed copula if it meets three conditions in addition to the obvious fact that its copula is in the purely grammatical present tense. First, (1) entails

(2) The Battle of Waterloo is (now) later than the Battle of Hastings,

where "is" is present-tensed, meaning what it does in "John is running." Second, (1) entails

(3) The Battle of Waterloo will always be later than the Battle of Hastings,

which contains the future tensed copula "will be" as modified by a temporal adverb (in this case, a universal temporal quantifier). Third, (1) is synonymous with a sentence (4) whose copula is the conjunction of the copula in (2) and the modified copula in (3):

(4) The Battle of Waterloo is, and will always be, later than the Battle of Hastings.

I do not hold that (1) entails a past tensed sentence that meets criterion C2, for reasons relating to the problem of future contingents, as I shall explain below.

Before I begin my argument, it is necessary to become clearer about the thesis for which I shall be arguing. Something needs to be said about the semantic content of the conjunctive tense of the copula in B-sentences. My argument will be that the conjunctive tense of the B-copula ascribes the property of *everlasting presentness*, that is, being present and being ever after present. The conjunctive tense ascribes this property not to the events that are B-related but to the events' *state of being B-related*. The events are said to be presently B-related and to be ever after B-related. The property of everlasting presentness is not an A-property, since A-properties are one of three sorts—(a) being present, (b) being presently past, and (c) being presently future—and everlasting presentness, although it includes two A-properties (i.e., being present and being presently future) is not an A-property. The thesis that B-sentences have a copula that is conjunctively present and future tensed entails that they ascribe both presentness and present futurity (the ascription of these two A-properties is part of the ascription of everlasting presentness), but it does not entail that they ascribe these two A-properties to the same item. If they did ascribe them to the same item, the B-sentence would be self-contradictory. Manifestly, "x is present, and x is future" is self-contradictory, as we know from our discussion of McTaggart's paradox. The conjunctive tense of the B-copula ascribes presentness to the events' state of being B-related and ascribes present futurity not to this state but to the state's *occupation* of future moments. Given this, the statement that everlasting presentness is ascribed to the events' state of being B-related can be misleading if left unanalyzed; for upon analysis, it is understood to mean that *one* of its parts (presentness) is ascribed to the event's state of being B-related, whereas another one of its part (present futurity) is ascribed to something else, namely, the state's occupation of future moments. I will explain the property of everlasting presentness in more detail in a moment, but first I wish to point out that the ascription of presentness and present futurity is consistent with point iii of my characterization of B-sentences; for point iii implies that B-sentences do not ascribe A-properties to the concrete events *to which the polyadic B-properties are ascribed* (e.g., the two battles) which is consistent with B-sentences' ascribing an A-property to something else, such as the two battles' state of being B-related.

An analysis of the property of everlastingness (= everlasting presentness) is best accomplished through first distinguishing the A-property of present futurity from the property of *continuing*. These properties can easily be confused, since they both can be expressed by "will be present." The difference is that something has the A-property of present futurity only if it is not present, whereas something has the property of continuing only if it is present. Tomorrow's sunrise is not present but is future. This entails that it will be present. But it is also true that John, who is present today, will be present tomorrow, assuming he does not die in the meantime. But it is mistake to assume that "will be present" has the same sense in

(5) The sunrise will be present

and

(6) John will be present.

In (5), "will be present" means "is future," such that "x is future" entails *that x now occupies a future moment, rather than a present moment*. But "will be present" in

(6) means "is continuing on into the future," such that "x is continuing on into the future" entails *that x now occupies the present moment and does not now occupy any future moment, but x will occupy some future moments when they become present.* Something x continues if there are several moments M1, M2, and M3, such that x occupies M1 and only M1 when M1 has presentness, occupies M2 and only M2 when M2 has presentness, and likewise for M3. This entails that if M2 is now present, then x is now occupying M2 and is not now occupying, but used to occupy, M1 and is not now occupying, but will occupy, M3.

The property of continuing, although different from the A-property of present futurity, includes this A-property as one of its parts. If x has the property of continuing, this means that (i) x has the A-property of presentness and (ii) x's *occupation of some immediately future moments* has the A-property of present futurity.

The property of continuing involves x's being present when *some* future moments become present, but the property of everlastingness involves x's being present when *each* future moment becomes present. Something x is everlasting only if, for each moment M that is later than the moment that is present, x will exemplify the property of occupying M and will exemplify this property when and only when M has presentness. Something x's property of everlastingness consists of (i) the presentness of x and (ii) the present futurity of x's *occupation of each future moment*. The component property ii can be stated more precisely in terms of degrees of futurity (such as being future by two minutes). Component ii should be understood as entailing that for each moment that is future to a certain degree, x's *occupation of that moment* is future to the same degree.

The future tensed copula, strictly speaking, ascribes only the A-property of present futurity; and if it appears in a sentence that ascribes continuing or everlastingness, that is because the future tensed copula is implicitly or explicitly modified by a temporal adverb. If "John will be present" and "John will be alive" ascribes *continuing*, rather than present futurity, that is because the "will be" is tacitly modified by the adverb "still," so that these sentences are synonymous with "John will still be present" and "John will still be alive." We may say, then, that the future tense in "will still be" and "will always be" ascribes the A-property of futurity but ascribes it only as a part of the more complex property of continuing or everlastingness that is ascribed by the entire phrases "will still be" and "will always be." We may say, accordingly, that if the copula in a B-sentence is present and future tensed, its present tense is unmodified, but its future tense is modified by the universal quantifier "always," such that the semantic content of the tense of the B-copula is made syntactically explicit by "is and will always be."

One last explanation of my thesis about the semantic content of B-sentences is needed before I present my arguments for it. The thesis for which I shall argue is *not* the traditional thesis of A-theorists like Gale (in his tenser period) that the B-adjectives "simultaneous," "earlier," and "later" can be explicated in terms of disjunctions of the three A-properties. It is part of the traditional thesis, for example, that "x and y are simultaneous" means "x and y either are both future to the same degree, or are both present, or are both past to the same degree." Whether or not this standard "reductionist analysis" of B-adjectives is sound is irrelevant to my argument, which is that B-sentences assert that the polyadic properties expressed by the B-adjectives are everlastingly exemplified by the relevant events.

As I have suggested, my argument concerns the *state* consisting of the exemplification of a B-relation by an event. An example of one of these states is the Battle of Hasting's *exemplification* of being earlier than the Battle of Waterloo. As I have mentioned before, I use "states" and "events" synonymously to refer to the exemplifications of n-adic properties by items, without pretending that this captures the ordinary use of "events" or "states." My thesis is that the states consisting of the exemplification of B-relations by events are ascribed everlasting presentness. This is proven by the following argument.

The states in question are either timeless or in time. They cannot be timeless, for there are no timeless states (as I shall demonstrate in section 6.5). If they are in time, they are either past, present, or future. An example will show that they cannot have the A-properties of present pastness or present futurity but can only have the A-property of presentness. Consider the sentence "The Battle of Hastings is earlier than the Battle of Waterloo." Let A be the Battle of Hastings and B the later battle. Then, there are several possibilities:

(7) A's standing to B in the relation of *earlier* has passed away (is past).

(8) A's standing to B in the relation of *earlier* is present.

(9) A's standing to B in the relation of *earlier* is still to come (is future).

Item (9) is obviously false, and the only question is whether (7) or (8) is true. It might be thought that (7) is true, since both A and B have passed away; that is, they now occupy a past moment, rather than the present moment. But

(10) The Battles of Hasting and Waterloo are no longer occurring

does not entail

(11) The two battles no longer possess any n-adic properties.

I argued in the last chapter that past events now possess some properties, such as the property of being past, occupying a past moment, and (perhaps) the properties of being remembered or referred to. Of course, it is also true that there are some properties that they used to possess but no longer do, such as the property of being witnessed by somebody or occupying this ground. The battles are now being referred to, but they are not now being witnessed and are not now occupying this ground. My thesis is that their B-properties are among the properties they presently possess. To suppose that their B-properties, like their being witnessed and occupying this ground, are no longer possessed, is to suppose, in part, that

(12) The Battle of Hastings is no longer earlier than the Battle of Waterloo,

which, insofar as it has a clear sense, is false. On one reading, (12) implies that the Battle of Hastings used to stand to the Battle of Waterloo in the B-relation of *being earlier* but now stands in some other B-relation to it, such as simultaneity or being later, which is false, since events cannot change their B-relations to other events. On another reading, (12) implies that the two battles used to stand in a B-relation but now stand in no B-relation. This is false because it suggests that the past is now temporally unordered, that nothing that has occurred is now earlier, later, or simultaneous with anything else that has occurred. It suggests that events are temporally ordered by the three B-relations only at the time that they are present, which is self-contradictory, since the B-relations of earlier and later obtain between two events only if at least one of the events is not present.

It might be suggested that (12) is now true, since events are B-related only if at least one of them is present. It would follow that the two battles are not now B-related. But this suggestion is also self-contradictory. According to this thesis, the Battle of Hastings is earlier than Waterloo if Waterloo is occurring; but once Waterloo is over, then Hastings is no longer earlier than it. Rather, it was earlier than Waterloo but is now neither earlier, nor later, nor simultaneous with Waterloo. This theory is also self-contradictory, although less obviously so. Consider that on this theory Waterloo is now (in 1989) 175 years earlier than an event E that occurs this year and is present, and Hastings is now 923 years earlier than E. This implies the contradiction that Hastings is now 923 years earlier than E but is not now earlier than something that is now 175 years earlier than E. If Hastings and Waterloo are now both B-related to E, that implies that Hastings sustains the triadic B-relation *( )* *being now 923 years earlier than (E), which is now 175 years later than (Waterloo),* which entails that Hastings sustains the dyadic relation to Waterloo of *now being 748 years earlier than* it.

This suggests that if two events are B-related, they are presently B-related, regardless of whether the events are past or present. The presentness of their B-relatedness is not fleeting, but everlasting; if A is presently earlier than B, then the proposition *that A is (present tense) earlier than B* will be true at each future time. But in deference to the problem of future contingents, I will not say that the presentness of the B-relatedness of events is sempiternal; that is, it is not the case that if two events are now B-related, then they not only always will be B-related but *always have been B-related;* for if they always have been B-related, then they were B-related when future. And if they were B-related when future, then they occupied future moments and future tensed statements about them had truth values, which is inconsistent with the assumption that regarding future contingents, the future is indeterminate and that future tensed statements about future contingents have no truth value. However, I have not argued that future tensed sentence-tokens about future contingents are truth-valueless, so I cannot claim here that I have demonstrated that the presentness of B-relatedness is everlasting but not sempiternal. Rather, I am making the considerably more modest claim that *if* the future tensed sentence-tokens about future contingents are truth-valueless, then the presentness of B-relations is everlasting and not sempiternal. I am assuming that B-states are everlasting for the sake of definiteness in the formulation of presentism and because it seems intuitively plausible to me that statements about future contingents are truth-valueless; but since this intuition is not backed by arguments against those who have contrary intuitions, this position remains, at least in this treatise, a mere assumption.

The argument I have presented for the multiple tensed character of the copulae in B-sentences is, in effect, a metaphysical argument based on the temporal status of the B-relatedness of events. The conclusion that B-sentences are everlastingly tensed could also have been reached by an argument based on the rules of use of B-sentences, for it is a rule that these sentences entail both present tensed sentences and future tensed sentences that are modified by "always." It is a rule that if I say, "The Battle of Waterloo is later than Hastings," I must also be disposed to assent to "The Battle of Waterloo is and always will be later than Hastings." But there is a stronger relation between these two sentences, for they are also *synonymous*. Their synonymity is suggested by the pertinent data about belief contexts. If the copula in B-sentences is present and future tensed, with its future tense being modified by a universal temporal quantifier, then this copula should be intersubstitutable *salva veritate* in belief contexts with more syntactically explicit expressions expressing the same semantic content, those of the form "is and always will be," where the first "is" is present-tensed. Consider

(13) John believes that Plato's birth is earlier than Aristotle's.

Truth value is preserved if we substitute

(14) John believes that Plato's birth is and always will be earlier than Aristotle's.

But different substitutions do not preserve truth value; for example, if we substitute for the "is" in (13) "was but no longer is," we get a false sentence. Only if John is mad, joking, or fails to understand the rules of language, would he believe that the embedded clause in (13) means or entails "Plato's birth was earlier than Aristotle's but no longer is." Likewise, if we substitute "at no time" we get a false sentence. To say that Plato's birth *at no time* is earlier than Aristotle's is tantamount to saying that it is false that Plato's birth is earlier than Aristotle's. I believe this is a rule of ordinary usage, although it is obviously not what many philosophers would mean by this sentence; for these philosophers would use this sentence in a technical sense to ascribe *timelessness* to these B-states. (L. Nathan Oaklander holds an atemporalist position about B-states roughly similar to this, for example.)[10] But I will argue in a later section that it is false that any state is timeless.

Just as the past tense of the copula in the A-sentence "John was running" refers thinly to a property of which presentness is a part, namely, present pastness, so the everlasting tense of the copula in the B-sentence "Plato's birth is earlier than Aristotle's" refers thinly to a property of which presentness is a part, namely, everlasting presentness. There is some distinct syntactical part of B-sentences, the tense of their copulae, that thinly refers to a thin property of which thin presentness is part, namely, the thin property of everlastingly presentness. It follows that presentness meets the criterion mentioned in section 6.1 for being a referent of all B-sentences. Presentness is also ascribed a monadic property by B-sentences, so that presentness meets the criterion for being a logical subject of these sentences. Presentness is ascribed a monadic property of the form *( ) inheres in the B-relatedness of the events E1 and E2 and always will inhere in their B-relatedness.* It follows from these considerations that ordinary B-sentences are translated by presentist sentences of a certain sort. "Plato's birth is earlier than Aristotle's" is translated by (expresses the same proposition as) the presentist sentence "Presentness inheres in, and always will inhere in, Plato's birth being earlier than Aristotle's." This presentist sentence makes syntactically explicit the reference to presentness and the indirect attribution to it of the monadic property of *inhering everlastingly in Plato's birth being earlier than Aristotle's birth.* Thus, if one wishes to token mentally, inscriptionally, or vocally presentist sentences, rather than ordinary sentences, for the purpose of making the presentist unity of reality explicit, all B-propositions would be expressed by sentences of the form "Presentness inheres everlastingly in E1's being B-related to E2." By tokening such sentences, one's mind becomes a mirror, as it were, of the unification of reality in the universal metaphysical subject.

6.4 Synthetic Generalizations

Synthetic generalizations include such sentences as

(1) All the people on the block are old.

(2) If anything is human, it is under nine feet tall.

(3) Being human requires having genes.

Most synthetic generalizations that occur as parts of ordinary usage are similar to (1) in being both tensed (i.e., singly tensed) and *de re.* The copula "are" in (1) is present-tensed; and the grammatical subject "All the people on the block" is used referentially to refer directly to the persons who, in fact, are on the block. Sentence (1) cannot be analyzed in the way familiar in textbooks on logic, as a material implication preceded by a universal quantifier, which has the logical form $(x) (Fx \supset Gx)$; for on this analysis, utterances of (1) would be true if there were no persons on the block. But these truth conditions do not belong to the rules of the ordinary use of (1). It is preposterous to think that someone could respond to an utterance of (1) with, "You are right, since there are no people on the block." Sentence (1) is uttered with truth if and only if there presently are some people on the block and each person on the block presently has the property of being old. A use of (1) expresses the *de re* proposition consisting of John, Alice, Fred, and Jane as propositionally E-related to the property of oldness, with presentness being C-related to the propositional complex consisting of John, Alice, Fred, and Jane as E-related to oldness. If the proposition is true, John, Alice, Fred, and Jane exemplify oldness, and their exemplification of oldness itself exemplifies presentness.

Since (1) is singly present-tensed, it is not an apparent exception to the theory that all sentences ascribe presentness or a property that includes presentness. The apparent exceptions are the "tenseless" sentences, those that do not have singly tensed copulae. My main concern in this section is with tenseless synthetic generalizations, such as (2) and (3). I use (2) to express a material implication preceded by a universal quantifier and (3) to express a law of nature. Both are synthetic generalizations; but (2) states an *accidental uniformity* and (3) a *necessary relation between universals*.

I shall begin by analyzing (2). It is doubtful whether any if/then sentence in ordinary language is correctly analyzed as a tenseless material implication preceded by a universal quantifier. It may well be the case that such sentences belong only to the artificial languages of some philosophical theories and twentieth-century logic textbooks. But there are true propositions expressible by such artificial if/then sentences, and these propositions are expressible by a different type of ordinary sentence. The proposition artificially expressed by (2) is expressible in ordinary language by the "tenseless" sentence

(2A) Nothing is both human and not under nine feet tall.

The quantifier "Nothing" expresses a second-order property, a property of the conjunctive property *being human and not under nine feet tall*. The property expressed by "Nothing" is *being unexemplified,* so that (2A) expresses the same proposition more perspicaciously expressed by the "tenseless" sentence

(2B) *Being human and not under nine feet tall* is unexemplified.

My thesis about universally quantified material conditionals is that their copula is "tenseless" in the sense of lacking a single tense but is conjunctively tensed, albeit with a species of conjunctive tense different from that of the copulae of B-sentences. The copula in (2B) is sempiternally tensed, for (2B) entails

(4) *Being human and not under nine feet tall* always was unexemplified

(5) *Being human and not under nine feet tall* is (now) unexemplified

and

(6) *Being human and not under nine feet tall* will always be unexemplified

and is synonymous with

(7) *Being human and not under nine feet tall* always was, is, and always will be unexemplified.

The copula in B-sentences is conjunctively tensed in that it is present-tensed and future-tensed, with its future tense being modified by "always." The copulae in empirical generalizations such as (2) have a different species of conjunctive tense; they are present-, past- and future-tensed, such that the past and future tenses are modified by "always." The evidence for this is that (2), (2A), and (2B) each entails (4)–(6) and is synonymous with (7). The entailments seem obvious; for (2B) cannot be true if, say, (4) is false. If (4) is false, then something is both human and nine-feet-or-more tall (where the "is" is "tenseless"), which contradicts (2B). There is also a synonymy between (2B) and (7); for "always was, is, and always will be unexemplified" is intersubstitutable with the "is unexemplified" in belief contexts; and the two sentences have a syntactico-semantic correlation.

That (2) ascribes sempiternally—or at least some temporal property—is also evinced by the following fact. If the purely grammatical present tense in (2) is not multiply tensed but is semantically a "timeless tense" (i.e., ascribes timelessness), then (2) would be synonymous with "if anything is-timelessly human, it is-timelessly under nine feet tall." But the "is" in (2) cannot mean this; for if anything is human and under nine feet tall, it is human and under nine feet tall in time and for a time, namely, during the time it lives and flourishes, and is not human and a certain height outside of time or at no time at all. This intuitive point can be backed up with an argument that the "is" cannot mean "is-timelessly," for there is an argument (to be presented in the next section) that there is no actual thing or state that is timeless.

I have not yet explained the parts and structure of the property of sempiternally or of the proposition expresses by (2B) and (7). This explanation will provide us with an understanding of the respect in which presentness is a *metaphysical subject* of the state of affairs corresponding to the proposition expressed by one of these sentences. This account will illustrate the presentist theory of the nature of the "general facts" that correspond to empirical generalizations. If an utterance of

(7) *Being human and not under nine feet tall* was always, is, and will always be unexemplified

is true, it corresponds to a state of affairs consisting of no substances or concrete events but solely of properties, namely, the property of *being human and not under nine feet tall*, the property of *being unexemplified*, and the property of *sempiternality*. This state of affairs consists of sempiternally as inhering in the *being unexemplified of being human and not under nine feet tall*. The proposition and corresponding state of affairs are isomorphic, so that the proposition consists of the relevant propositional parts as E-related or C-related. As we know, an item A is propositionally E-related to an item B if A *exemplifies* B if the proposition is true. Now, *being unexemplified* is an indirectly possessed higher-order monadic property, such that the proposition expressed by a use of (7) expresses

(P1) *Being human and not*   is-E-related-to   *being unexemplified.*
      under nine feet tall

The property of sempiternality, on the other hand, is propositionally C-related to (P1), so that

(P2) *Being sempiternal*   is-C-related-to   the *being unexemplified of being human and not under nine feet tall.*

Item (P2) is such that if the proposition is true, the property of sempiternality is exemplified by whatever quasi-corresponds to the propositional complex (P1). What quasi-corresponds to this complex is an abstract event or state, the state consisting of the exemplification of *being unexemplified* by *being human and nine-feet-or-more tall.*

This analysis suggests that contrary to initial appearances or some unsubstantiated dogmas, propositions expressed by universally quantified material implications are structurally isomorphic to singular *de re* propositions. Compare (P1) and (P2) to the two propositional complexes in the proposition expressed by an utterance of "John is running" (as analyzed in section 6.1):

    (P3) John   is E-related-to   running.

    (P4) Presentness   is C-related-to   the running of John.

Here, John has an analogous propositional role to *being human and nine feet or more tall,* running has a role analogous to *being unexemplified,* and presentness a role analogous to *being sempiternal.*

Sempiternality differs from everlastingness in that the former is expressible by "always was, is, and will always be," whereas the latter is expressed merely by the present and future tensed part of this phrase—"is and will always be"—so that it does not consist of the pastness-involving part of sempiternality. But what, *exactly,* is the structure and composition of sempiternality? I analyzed a part of sempiternality, namely, everlastingness, in the last section, in terms of two properties: *presentness* and the *futurity of something's occupation of each future moment.* But this analysis includes the quantifier phrase "each future moment," which is susceptible of further analysis. Let us call the abstract state to which sempiternality is ascribed by (7) the state S. The state S is ascribed sempiternality only if it is ascribed presentness and a complex futurity-involving property that consists of these parts:

    F1 Being a presently future moment

    F2 Will be unoccupied by [S] when it becomes present

    F3 Being unexemplified.

These parts are ordered in such a way that F3 is ascribed to the conjunction of F1 and F2; that is, *being unexemplified* is ascribed to the property of *being a future moment and will be unoccupied by [S] when it [the moment] becomes present.* The ascription of the polyadic property of sempiternality to a state S involves ascribing the conjunctive property *F1 and F2* to S (in a way that is expressed by placing S in brackets in F2) and ascribing *being unexemplified* to the conjunctive property *F1 and F2.* The same holds for the pastness involving property, so that if the ascription of sempiternality to S belongs to a true proposition, this ascription (which is the propositional complex P2 represented above) quasi-corresponds to this complex state:

(a) *Presentness* as inhering in *S*.

(b) *Being unexemplified* as inhering in *being a future moment and will be unoccupied by S when it [the moment] becomes present.*

(c) *Being unexemplified* as inhering in *being a past moment and having been unoccupied by S when it [the moment] was present.*

The fact that (a) belongs to this state is sufficient to guarantee that the complex state of affairs to which the proposition expressed by the aforementioned (2B) corresponds has presentness for a metaphysical subject.

The subject of this section is synthetic generalizations, which includes both statements of accidental uniformities and statements of laws of nature. That all humans are under nine feet tall is an accidental uniformity, since it is merely an accident that no human has been, is, or will be nine-feet-or-more tall (assuming that this is, indeed, a fact). It is nomologically consistent with human nature to be nine feet or nine-feet-and-one-inch tall. Laws of nature are different, but their exact nature remains a subject of controversy. The most widely held theory of natural laws in this century is the so-called regularity theory. This theory has many versions, but most of them involve the idea that laws of nature are expressed by universally quantified material implications, with the provisos that they (i) either involve existential quantification (and are of the form $[\exists x]\ [Fx]\ \&\ [x]\ [Fx \supset Gx]$) or are deducible from laws that do involve existential quantification; (ii) do not involve reference to particular individuals; and (iii) entail some counterfactuals and have simplicity in some respect (e.g., are theorems of each true deductive system that best combines simplicity and strength). Recently, however, a new theory of natural laws has been developed by D. M. Armstrong, Michael Tooley, and Fred Dretske.[11] According to them, laws are relations of nomic necessitation among universals and are expressed by sentences of such forms as *"Being an F* necessitates *being a G"*[12] or "N(F, G)," where N is the relation of nomic necessitation. According to Armstrong, Tooley, and Dretske, the relation of nomic necessitation is a logically contingent relation. But there is a third view of laws of nature that is similar to the Armstrong–Tooley–Dretske account except for the fact that the relation N is a species of logical necessity, namely, broadly logical necessity, or "metaphysical necessity" (as it is usually called). This third theory may be characterized as a version of the essentialist theory of laws of nature. (At least one philosopher has argued for this third theory, namely, Alfred J. Freddoso.)[13] I shall not attempt to decide which of these three theories is true but shall confine myself to showing how the third theory should be interpreted by a presentist.

There is a threefold distinction to be made among the sentence; the proposition that states the law of nature; and the law of nature itself, which is the state of affairs that corresponds to the proposition. Consider the metaphysically necessary sentence "All humans have genes." This expresses a proposition that includes among its parts the property *humanity* and the property of *partly composed of genes*, and its states that these two properties stand in a relation of metaphysical necessitation. This law of nature is most perspicuously stated by the "tenseless" sentence

(8) *Being human* metaphysically necessitates *having genes*.

I believe the verb "necessitates" in (8) is sempiternally tensed, since (8) entails

(9) *Being human* has always metaphysically necessitated *having genes*.

(10) *Being human* metaphysically necessitates (present tense) *having genes*.

(11) *Being human* will always metaphysically necessitate *having genes*.

Sentence (8) is synonymous with a sentence that includes the conjunction of the tensed copula and verbs of (9)–(11).

In support of this contention, let us draw the consequences of the supposition that (8) does not entail (9). Then it is possibly the case that (8) is true and yet that being human has not always metaphysically necessitated having genes. If it has not always metaphysically necessitated this, then at some time in the past, *being human* did not stand to *having genes* in the relation of metaphysical necessitation; that is, it was then not a law of nature that all humans have genes. But this contradicts the standard definition of a law of nature, namely, that if L is a law of nature, then it is false that L obtains at some times and not at others.

It might be objected that laws of nature obtain timelessly and that for this reason, *being human* has not always metaphysically necessitated *having genes*. This objection will be answered in the next section, in which I argue that nothing is timeless. But if we grant that nothing is timeless, it follows that natural laws obtain sempiternally; for the alternative is the falsity that they obtain at some times and not at others. We may say that a law of nature of the form N(F, G) is a state of affairs consisting of presentness as inhering in *being F's* standing in the relation of metaphysical necessitation to *being G,* of presentness as having always inhered in their relatedness, and of presentness' being such that it will always inhere in their relatedness. In this manner, presentness is a metaphysical subject of each of the states of affairs to know which is a goal of theoretical science. Theoretical science is about presentness in respect of some its sempiternal properties, namely, the sempiternal properties that are composed of the relevant metaphysically related universals. Laws of nature are expressible in presentist sentences of such forms as "Presentness inheres sempiternally in the metaphysical necessitation of *being F* by *being G.*"

6.5 Tautological and Analytic Sentences

I shall use the words *tautological truth* and *analytic truth* in this section to refer to sentences. A tautology (as I shall here use the phrase) is a principle of logic (including at least propositional logic and first-order quantification theory) or a substitution instance of such a principle. An example of such a principle is "If no individuals that have a certain property also have a certain second property, then no individuals that have this second property also have the first property." A substitution instance of this principle is "If no trees are animals, then no animals are trees." An analytic truth is any sentence that is reducible to a tautology by a substitution of a synonym for some words or phrases in the sentence. "A bachelor is an unmarried man" is analytic, since it is reducible to a tautology by substituting "a bachelor" for "an unmarried man." Mathematical truths are sometimes regarded as synthetic a priori, but I shall here assume the different and more predominant view that they are analytic.

My concern in this section will be with principles of logic and mathematics, since these are the tautological and analytic sentences that seem to be the most plausible candidates for sentences that do not have presentness for a logical subject. It is clear that many substitution instances of these principles have a temporal content (e.g., "It is false that John was both running and not running at noon yesterday"); consequently, they pose less of a threat to the presentist. My aim, accordingly, is to show that principles of logic and mathematics have a temporal content and, specifically, that they have conjunctively tensed copulae or verbs that serve to ascribe the property of sempiternality to the relevant logical or mathematical states. In this respect, the semantic content of these sentences is similar to that of synthetic generalizations.

The argument that principles of logic and mathematics are sempiternally tensed has harder-going than the argument that synthetic generalizations are. In the last section, I stated that synthetic generalizations entailed singly tensed sentences, for example, that "All humans have genes" entails "If anything is now a human, it now has genes." I did not provide much by way of supporting argument for these entailments, since it is both obvious and widely accepted that there are such entailments. The reason for the obviousness of the entailments is that synthetic generalizations (at least of the sort I discussed) are generalizations about sorts of things that obviously exist in time if they exist at all. But the case is different with logical and mathematical sentences, for it is not obvious that the abstract objects they are about exist in time. Indeed, it seems obvious to many philosophers that these objects do not exist in time. Consequently, arguments for the relevant tensed entailments and synonymies must be developed at greater length than in the last section.

I being with the claims that the "tenseless" mathematical sentence

(1) Two equals one plus one

entails the present tensed sentence

(2) Two (now) equals one plus one.

Now it seems to me evident that (1) does entail (2) and that this is built into the rules of use of (1). For example, someone might say, "Two equals one plus one and that implies, of course, that two *now* equals one plus one"; and his addressee would respond, "Of course." I believe, furthermore, that not only do these entailments obtain but also a synonymy, namely, that the "tenseless" sentence "Two equals one plus one" is synonymous with

(3) Two always equals one plus one

or with the more syntactically explicit

> (4) Two always has equaled one plus one, (now) equals one plus one, and
> always will equal one plus one.

Sentences (3) and (4) are both intersubstitutable *salva veritate* in belief contexts
with the "tenseless" sentence (1) except in the abnormal cases where certain (arguably false) beliefs of philosophers are being reported. There is also a syntactico-
semantic correlation between (1) and (4), for the semantic content of the purely
grammatical present tense of the "equals" in (1) can be correlated with the semantic
content of the complex grammatical tense of "always has equaled, . . . (now)
equals, . . . and always will equal."

But these entailments and synonymies are frequently denied. P. F. Strawson, for
example, expresses a widely shared belief when he avers:

> The sentence "There is at least one prime number between 16 and 20" does not
> mean that there is one at the moment. Nor does it mean that there always has been,
> is now, and always will be one such prime number. For it makes no sense to speak
> of such a number beginning or ending or going on and on. We might say that in this
> context "There is . . ." is timeless.[14]

Strawson's argument for the timelessness of numbers is invalid, for "Numbers
are sempiternal" does not entail "We can speak sensibly of numbers beginning or
ending or going on and on." If a number is sempiternal, there is no time earlier than
all the times at which the number exists and no time later than all the times at which
the number exists; and this is necessarily the case. Numbers are necessary existents
and necessarily exist at every time. It is for this reason that it makes no sense to
speak of a number beginning or ending. (But if time begins and ends, then it makes
sense *in this context* to speak of numbers beginning and ending, namely, at the first
and last moments of time.) Strawson's other remark about "going on and on" seems
to be a remark about the ordinary use of "going on and on," and he is right that this
phrase is not "correctly" applied to numbers. But the reason for this is not that
numbers are timeless but that they are not processes, and it is only processes that are
correctly said to "go on and on." It is not correctly said of a substance, such as John,
that "he is going on and on"; and so if Strawson's argument were valid, it would
follow that all substances, including John, exist timelessly, which is absurd. If we
do say "John is going on and on," what we mean is that some process that John is
sustaining, such as a speech, is going on and on. On the other hand, if "goes on and
on" is used in a nonordinary and technical sense to mean "perdures," then it does
make sense to speak of numbers going on and on or perduring forever.

A more fundamental argument for the timelessness of numbers and the universals referred to by logical principles is presented by Panayot Butchvarov. In his
Concept of Knowledge, Butchvarov makes some penetrating criticisms of some of
the standard logico-linguistic definitions of necessary truth, but he believes that this
notion is definable in terms of the thesis that the subject matter of necessarily true
propositions is timeless. One of his main arguments for the timelessness of their
subject matter is that these propositions cannot be prefixed by temporal operators.

He writes" "I have said that a nontemporal proposition is one that cannot be intelligibly prefixed with a date. For example, 'Three and two are five' is a nontemporal proposition because it would be nonsense to say that today three and two are five."[15] J.J.C. Smart makes a similar claim: "When we say two plus two *equals* four we do not mean that two plus two equals four at the present moment. Nor do we mean that two plus two always equalled four in the past, equals four now, and will always equal four in the future. This would imply that two plus two will equal four at midnight tonight, which has no clear sense."[16] I would admit that in some contexts utterances of the sort that Butchvarov and Smart allude to have no clear sense. If I say, out of the blue, "Two plus two will equal four at midnight tonight," my statement will have no clear sense or will seem unintelligible. But this is not because this sentence is analytically false or expresses a "category mistake" but because in some contexts an utterance of this sentence *conversationally implicates* (in Grice's sense)[17] sentences that are analytically false. An utterance of a sentence S_1 conversationally implicates another sentence S_2 if (i) S_1 does not entail S_2; (ii) the pragmatic maxims governing discourse (such as "do not give less (relevant) information than you are capable of giving easily") indicate that the person would not have uttered S_1 unless she also believed S_2; and (iii) the suggestion of S_2 by the utterance of S_1 (by virtue of the pragmatic maxims) is not built into, or derivable from, the conventional meanings of S_1. Grice distinguishes between general and particular conversational implicatures; and the implicature pertinent to our example is general, that is, it is an implicature normally carried (i.e., in the absence of special circumstances) by the utterance of the sentence in question. Now the utterance of a sentence of the form "x will stand in the relation R to y at midnight tonight" conversationally implicates sentences of the form (a) "x stands in the relation R to y at some times and not at others" and (b) "It might not have been the case that x will stand in R to y at midnight tonight." If a mathematical sentence such as "Four will equal two plus two at midnight tonight" is uttered by itself, in the absence of special circumstances, then it conversationally implicates the analytically false sentences "Four equals two plus two at some times and not at others" and "It might not have been the case four will equal two plus two at midnight tonight." It is because of these conversational implicatures that the utterance of "Two plus two will equal four at midnight tonight" has no clear sense. But if a sentence of the form "x will stand in the relation R to y at midnight tonight" is uttered *in special circumstances,* these conversational implicatures are canceled. For example, if a sentence of this form is uttered as a deduction from the relevant modal and temporal sentences, the implicatures are canceled. If I say "x necessarily and always stands in the relation R to y; therefore, x will stand in R to y at midnight tonight" then the implicatures (a) and (b) are canceled. If uttered in a circumstance of this form, the sentence "Two plus two will equal four at midnight tonight" has a clear sense. Thus, I believe the "unintelligibility" and "unclear sense" that Butchvarov and Smart allude to are not semantic properties of the sample sentences they introduce but are pragmatic properties that they possess in some circumstances of utterance. The propositions they express are intelligible and have a clear sense and can be clearly comprehended if the sentences are uttered in the appropriate circumstances.

The proponent of the timeless content of mathematical propositions might try and turn the tables and argue that sentences such as "Two now equals one plus one" are themselves merely conversationally or conventionally implicated by the tenseless mathematical sentences and therefore that my claim for an entailment relation is false. Since they are implicated and not entailed, the proponent may continue, the timelessness of these propositions can still be maintained.

This counterargument fails because it can be shown that the relation between the sentences is one of *entailment,* not Gricean implicature. One sentence S_1 entails another S_2 if there is no possible circumstance in which S_1 could be uttered with truth and S_2 with falsity. Since there is no possible circumstance in which "Two (now) equals one plus one" could be uttered with falsity, it follows that it is entailed by the "tenseless" sentence (1).

A different line of attack is pressed by W. Kneale:

> An assertion such as "There is a prime number between five and ten" can never be countered sensibly by the remark "You are out of date: things have altered recently." And this is the reason why the entities discussed in mathematics can properly be said to have a timeless existence. To say only that they have a sempiternal or omnitemporal existence (i.e. an existence at all times) would be unsatisfactory because this way of talking might suggest that it is at least conceivable they should at some time cease to exist, and that is an absurdity we want to exclude.[18]

I would respond that at best, the assertion that they have a sempiternal existence, if uttered in certain contexts, may carry the conversational implicature that it is possible these entities should at some time cease to exist. But if uttered in the appropriate context, this is not implicated; more importantly, it is not *entailed* by the assertion that they exist sempiternally. A sentence of the form "X exists at all times" does not entail a sentence of the form "It is possible that x ceases to exist, such that there may be some times later than all the times occupied by x," for "x exists at all times" is logically consistent with "x necessarily exists at all times." This is a point I made earlier against Strawson. Kneale might respond that the thesis of timeless existence must be introduced to explain why "There is a prime number between five and ten" can never be countered sensibly by the remark "You are out of date: things have altered recently." But there is another explanation, namely, that it is logically impossible for this mathematical state of affairs to alter; if there is a prime number between five and ten at one time, then it is logically necessary that there is such a number at every time.

Butchvarov advances a more sophisticated version of this argument in *The Concept of Knowledge.* He writes:

> A necessarily true proposition, we can now say, is one that describes what is immutable, incapable of ever changing, in itself and not because of causes. . . . How is it possible for something to be immutable and yet not because of causes? . . . The answer . . . is that something can be immutable, and yet not because of causes, in the sense that it is not in time at all. . . . The notion of noncausal immutability is . . . simply the descriptive notion of the nontemporal.[19]

As Butchvarov acknowledges, if something is immutable, it can never change. If something is incapable of change because of causes, then it is incapable of change because causes for its changing are never present or because causes for its not changing are always present.[20] Butchvarov passes from these plausible observations to the conclusion that immutability that is not due to causes is due to timelessness, but this does not follow. Admittedly, if something is timeless, it is noncausally immutable; but being noncausally immutable does not entail being timeless. There is no contradiction in the supposition that "the state of affairs S obtains at all times and cannot fail to obtain at any time and its immutability is not because of causes but because of the nature of S itself." The reason Butchvarov argues for the timelessness of the states of affairs that correspond to necessary truths is that he is endeavoring to give a reductive analysis of "necessary truth," a definition that does not itself include the modal notion of necessity. But we are not forced to offer a reductive definition of necessary truth. Some notions are primitive and that of necessary truth is as good a candidate for a primitive notion as anything. But if it is insisted that it be defined, we could define it in a way that does not include the notion of necessity but does include sempiternality. We could say that a necessarily true proposition is one that corresponds to a state of affairs that obtains *at all moments in all possible worlds.* (In timeless worlds there is a single moment, the "standing now"; and in these worlds, the states of affairs in question obtain at this single moment. Timeless worlds will be further discussed later.)

I shall now pass to a second stage of my argument for the sempiternally tensed character of logical and mathematical principles. This second stage is strictly metaphysical in the sense that it is not about the entailment and synonymy relations among sentences but about features of the world as a whole. The supposition that there are timeless states of affairs corresponding to logical and mathematical sentences is inconsistent with the nature of the actual world. In the actual world, nothing exists timelessly, and no state obtains timelessly. This can be proven if a certain definition of temporal existence is adopted, namely,

D1 X exists in time = df. x exists in time if and only if there is some time t at which x possesses some monadic property F and some different time t' at which x does not possess F.

By "monadic property" I mean both nonrelational monadic properties, such as *being red,* and relational monadic properties, such as *being referred to by John.* Definition D1 entails that something exists in time if and only if some of the states of x succeed one another, that is, if x is successively in different states. A state is an exemplification of a property by something; and if x possesses F at t and does not possess F at t', then x's state of being F is succeeded by its state of being not F.

Now it might be argued that mathematical and logical states of affairs exist in time, since they successively exemplify such relational monadic properties of *occupying moment M1, which is present; occupying moment M2, which is present;* and so on. But such an argument is obviously question-begging and will convince no atemporalists. What is needed is a premise that temporalists and atemporalists alike will grant, such as that mathematical and logical states of affairs can be referred to. Let us use a premise such as this to construct an argument that nothing exists timelessly.

(4) For any x that is alleged to be timeless, x is now the referent of my present utterance U of "x is timeless."

(5) Therefore, x now possesses the relational monadic property of *being a referent of U*.

(6) But now, a moment later, U has passed away and is not presently referring to x.

(7) Therefore, x no longer possesses the property of *being a referent of U*.

(8) Therefore, by D1, x exists in time.

A similar argument could be stated in terms of mental or intentional reference to x, so that (4) would read, in part, "John is now (mentally) referring to x" and (6) would read, in part, "John is no longer referring to x."

Let x range over all mathematical and logical propositions and states of affairs. It follows from the application of D1 and (4)–(8) that all such propositions and states of affairs exist in time.

Of course, there has been a long tradition from Plato onward that propositions, universals, and abstract objects "exist timelessly." But they tend to mean by this something that I would agree with but would express differently, for example, that propositions and universals do not move (recall that Aristotle used "time" to express the property of *being the number of a motion*), or do not come to be and pass away, or do not mature or decay, or do not engage in causal interactions with other items, or do not emit or receive light signals, or the like. Something can acquire and lose relational properties of *being referred to,* yet not move, come to be or cease to be, mature or decay, engage in causal interactions, or emit and receive light signals. Thus, if logical and mathematical states of affairs are "in time" in the sense of D1, that is consistent with their being "outside of time" in one of these other senses.

It might be thought that D1 does not reflect a legitimate use of the phrase "exists in time." But such an objection is untenable. This phrase is used in a legitimate sense if it is predicated of some item x whose complete description involves the ineliminable use of temporal predicates. An item is completely described if all its states are described, that is, if for every property F it possesses, it is described as possessing F. Now, "successive," "earlier," and "later" are temporal predicates; and they are ineliminably involved in the complete description of an item x if they are required in the description of some of x's states. These predicates are required in the complete description of logico-mathematical phenomena, since they are needed to describe their states of being referred to by linguistic and mental occurrences. For any familiar logical or mathematical object (e.g., triangularity), we must include in its complete description such facts as that it exemplified *being referred to by Plato* before it exemplified *being referred to by Descartes;* and in this respect, temporal predicates such as "before" are ineliminably involved in its complete description. Accordingly, there is a legitimate sense of "exists in time" that can be applied to these objects.

There are narrower and wider legitimate senses of "exists in time," and these senses may be correlated with different degrees of completeness of the description of something. In the broadest or most latitudinarian sense, something "exists in time" if and only if temporal predicates are required to describe all the states of the item. In a narrower sense of "exists in time," something exists in time if and only if temporal predicates are required to describe its nonrelational states, that is, the item's possession of nonrelational monadic properties. In a still narrower sense, something exists in time if and only if temporal predicates are required to describe its monadic states that are terms of causal relations, such that if none of its monadic states that are terms of causal relations it does not exist in time in this sense. In a narrower sense still, something exists in time if temporal predicates are required to describe its monadic and causal states that are terms of relations of light-connectability, such that it does not exist in time in this sense if none of its states are light-connectable to other states. Many disputes about what does, and what does not, exist in time may be resolved by specifying which of these or other senses of "exists in time" is being employed. Now it may be that D1 articulates that broadest legitimate sense of "exists in time"; but that it articulates *some* legitimate sense, wide or narrow, is all I need for my argument about the sempiternality of logical and mathematical states of affairs (as I shall indicate shortly).

But first let me further attest to the legitimacy of my use of "exists in time." Some evidence for this consists in the fact that philosophers apprised of this usage often admit that in this sense of "exists in time," it can be said with truth that all abstract objects exist in time. A case in point is Frege, who throughout his life was a vociferous champion of the timelessness of propositions ("thoughts," in his terminology). In 1918–19 he conceded: "By apprehending a thought I come into a relation to it and it to me. It is possible that the same thought that is thought by me today was not thought by me yesterday. In this way the strict timelessness [of thoughts] is of course annulled."[21]

This broad use of "exists in time" is also adopted by Nicholas Wolterstorff. Wolterstorff uses "aspect" to mean something's actually having an n-adic property and writes: "*3's being referred to by Euclid* was an aspect of the number 3, and *3's being referred to by Cantor* was another aspect thereof. And of course the former preceded the latter. So, by our definition, 3 is not eternal [timeless]."[22] The "definition" Wolterstorff refers to is "x is eternal if and only if x has no aspect which is a member of the temporal array." The temporal array is the set of all B-related aspects; and some aspect is a member of this array if it is earlier than, later than, or simultaneous with, another aspect. Thus, Wolterstorff is implying that the number 3 exists in time, since its *being referred to by Euclid* is earlier than its *being referred to by Cantor,* which conforms to the sense of "exists in time" articulated in D1. Wolterstorff defends his definition of eternity against an objection that may also be made against D1; and I shall quote Wolterstorff's defense at length, since I believe it to be sound:

> [The objection is that] the fact that something is successively referred to should not be regarded as ground for concluding that it is not eternal. For after all, successive references to something do not produce any change in it. Although they produce variation among its aspects, they do not produce a changeful variation among them.
>
> In response to this protest it must be emphasized that the concept of an eternal being is not identical with the concept of an unchanging being. The root idea behind the concept of an eternal being is not that of one which does not change but

rather that of one which is outside of time. And a question of substance is whether an unchanging being may fail to be eternal. The most thoroughgoing and radical way possible for an entity to be outside of time is that which something enjoys if it satisfies our definition of "eternal." And it must simply be acknowledged that if an entity is successively referred to, then it is not in the most thorough-going way outside of time. There is temporal succession among its aspects.[23]

Delmas Lewis' uses a similar criterion for temporal existence; for he argues that if God is observed by P at t and is not observed by P at t′, then God must be in time, since God acquires and loses properties of being observed by P.[24]

Now if we assume the results of previous chapters, namely, that the tensed theory of time is true, the following principle is analytically true:

P1 If something x exists in time (in any of the legitimate senses), then x is either past, present, or future.

I think it is obvious that if the tensed theory of time is true, then "x exists in time but is neither past, present, nor future" is self-contradictory. Accordingly, since D1 articulates a legitimate sense of existing in time and since all logical and mathematical propositions and states of affairs exist in time in the sense of D1, it follows that all these propositions and states of affairs are either past, present, or future. If they are past, present, or future, then they are present; for logical and mathematical propositions and states of affairs are necessarily present at each present time if they exist in time at all. It can never be the case that they are passed away or are still to come, but it must always be the case that they are present—which is to say that they are necessarily sempiternal.

I think there is also a phenomenological argument that logical and mathematical propositions, states of affairs, and the like are temporally present. Right now, I am thinking of the property of triangularity. It intuitively seems to me that triangularity, which is now before my thoughts, is existing now, that it is existing simultaneously with my thinking of it. It seems that it is present and is not passed away or future and that it will remain present after I cease to think of it. Given that there are no phenomenological intuitions that are incompatible with this intuition, it follows that this intuition is veridical. (At least, it follows if one accepts certain criteria of phenomenological or intuitional truth.)[25]

Given that these several arguments are sound, it follows that logical and mathematical states of affairs exist sempiternally. This entails that the theory that the purely grammatical present tense of the copulae or verbs in true mathematical and logical sentences cannot be interpreted in the manner advocated by Strawson, Butchvarov, and Smart as a "timeless tense," that is, as ascribing timelessness to the states of affairs, but should be interpreted as conjunctively tensed—specifically, as sempiternally tensed.

The theory I am outlining entails that there is a distinction between the mathematical or logical sentence, the proposition expressed by the sentence, and the state of affairs that corresponds to the proposition. But are there mathematical and logical states of affairs? According to some theories, logical and mathematical sentences are true not by corresponding to necessarily obtaining states of affairs but by virtue of the definitions of the words occurring in them, or by virtue of their logical or syntactical form, or by virtue of the rules of our language, or the like. I believe that these theories are false and would refer the reader to Butchvarov's *Concept of Knowledge* for some sound criticisms of these and other "logico-linguistic theories" and for a defense of the view that these sentences are true by virtue of corresponding to logical or mathematical states of affairs. But I shall here briefly present an additional argument for the "correspondence theory" of the truth of these sentences. This argument requires the plausible assumption that these sentences express true propositions. A use of "Two equals one plus one" expresses a proposition that consists of two, equality, one, plus, and sempiternality. (We may assume here the theory, of, e.g., Armstrong and George Bealer, that numbers are properties.)[26] These various n-adic properties are propositionally related, such that *two* is E-related to *being equal to one plus one* and *sempiternality* is C-related to *two's being equal to one plus one*. If the proposition is true, then two exemplifies the relational property of being equal to one plus one and its exemplification of this property itself exemplifies sempiternality. But the proposition includes two only as propositionally E-related to one plus one. It does not include two's state of *exemplifying* this relational property. This state belongs to the state of affairs that corresponds to this proposition. This state of affairs necessarily obtains, that is, obtains at all moments in all possible worlds, which explains why the proposition is true at all moments in all possible worlds. If one denies that there are such states of affairs, one is committed to the view that mathematical entities are propositionally related to one another but do not stand in mathematical relations (e.g., of equality) to one another, which implies that mathematical propositions are one and all false.

I shall close this section by considering an objection to my theory of the sempiternality of logical and mathematical states of affairs that is based on the possibility of eternal worlds, worlds in which there is no time. The objection is that logical and mathematical propositions are true in every possible world but that they cannot be true in eternal worlds if they ascribe the property of sempiternality.

This objection can be refuted once we define the notion of an eternal world. I have elsewhere criticized the traditional Boethian theory of eternity[27] and shall here engage only in the positive task of constructing an alternate definition of eternity. As I use these expressions, "eternal" means "timeless" and "eternity/timelessness" means "eternal presentness/timeless presentness." The distinction that needs to be defined is that between *eternal presentness* and *temporal presentness*.

We begin with the concept of temporal presentness and abstract certain features until a concept of eternal presentness is reached. The difference between temporal and eternal presentness is a difference in the manner in which presentness inheres in something. Presentness inheres in an existent E in a *temporal manner* if and only if presentness inheres in E and either (i) there is some future or past time at which presentness does not inhere in E or (ii) there is some future or past time at which presentness does inhere in E. Condition i applies to transiently present existents only, to existents that are present at some times but not others. Condition ii applies to transient existents that are present for two instants or longer and to sempiternal existents, ones that are present at every instant of time. The "or" in the definition is inclusive, which means that conditions i and ii may both apply to some existents, that is, to all transient existents present for two instants or longer. Presentness inheres in an existent E in an *eternal manner* if and only if it inheres in E but not in a temporal manner; that is, presentness inheres in E eternally if and only if present-ness inheres in E and it is not the case that (i) there is some future or past time at which presentness does not inhere in E and it is not the case that (ii) there is some future or past time at which presentness does inhere in E. Suppose that one of these conditions does obtain. If presentness inheres in E and there is some past or future time at which it does not inhere in E, then it is true of E either that *it is present but was not present awhile ago* or that *it is present but eventually will not be present;* and in either case, E is present transiently and therefore is not eternal. But if presentness inheres in E and there is some past or future time at which it also inheres in E, then it is true of E either *that E is present and has been present* or that *E is present and will be present;* and in either case, E is present at several times and thus does not exist eternally. But if E is present and there is no past or future time at which E is either present or not present, then E is present *pastlessly* and *futurelessly* and thus is present *timelessly.* Such timeless presentness is what I call eternal presentness.

It is evident from these considerations that if something is eternally present, then nothing is temporally present and that if something is temporally present, then nothing is eternally present. Something is eternally present only if there is no time, but that does not imply that in eternity there are no properties of degrees of pastness and futurity. It implies merely that these properties are unexemplified. These proper-ties are themselves eternally present in eternity, such that every degree of pastness and futurity is present pastlessly and futurelessly.

We are now in a position to respond to the objection that propositions cannot be true in eternal worlds if they ascribe the property of sempiternality. My response is that this objection is based on a misunderstanding of the property of sempiternality, that is, the property of *being present and of always having been present and always going to be present.* "Always" is a universal quantifier over times; it means *for any time t.* Now universal quantification is devoid of existential import; it does not imply that there exist any values of the variable that is bound by the universal quantifier. For example, universal quantification over past times does not imply there are any past times. "E always has been present" means "For any x, if x is a past time, then E was present when x was present." It is false that something has always been present if and only if there is some past time that is unoccupied by the item in question. But if there are no past times, then it is not false that something E has always been present. Indeed, it is then true; for "E has always been present" is true just in case the conjunction of (a) *there is some past time x* and (b) *x is not occupied by E* is false. Since propositions a and b are not both true in eternal worlds, it is true in these worlds that E always has been present. Accordingly, if logical and mathematical propositions ascribe sempiternality, they are true in all eternal worlds, as well as in all temporal worlds. This entitles us to the conclusion that logical and mathematical sentences are synonymous with presentist sentences, that "One plus one equals two" is synonymous with "Presentness inheres sempiternally in the equality of one plus one to two."

6.6 Non-English Sentences

If the arguments in the foregoing sections are sound, I have established that presentness is a logical subject of every *English* A-sentence, singular and existential sentence, B-sentence, synthetic generalization, and tautological and analytic sentence. I have also argued that presentness is a metaphysical subject of every state of affairs that corresponds to a true token of one of these sentences. But these facts do not entail the thesis of presentism, that presentness is a logical subject of every possibly true sentence in any possible language, and is a metaphysical subject of every state of affairs. In order to establish this, it must be shown that every *non-English* sentence has presentness for a logical subject. This can be shown if we connect presentism with some of the ideas in transformational grammar.

A central idea of standard transformational grammar is that there are substantive and formal *linguistic universals,* which are phonological, syntactical, and semantic features that are common to all natural languages. The arguments for presentism may be construed as arguments for the existence of certain substantive universals. In *Aspects of the Theory of Syntax,* Noam Chomsky distinguishes three types of substantive universals, phonological, syntactical and semantic. Substantive phonological universals concern the acoustic features of speech common to all languages and substantive syntactical universals include such common features as the noun/verb distinction. Chomsky says this about substantive semantic universals:

> A theory of substantive semantic universals might hold for example, that certain designative functions must be carried out in a specified way in each language. Thus it might assert that each language will contain terms that designate persons or lexical items referring to certain specific kinds of objects, feelings, behavior, and so on.[28]

Presentism may be understood as the argument for a certain sort of substantive semantic universal, that each language is such that every sentence in it has presentness for a logical subject. If we combine this with the substantive syntactic universal that every sentence has a noun phrase/verb phrase distinction in its underlying syntactical structure, then presentism may be understood as the theory that every sentence in every language is such that the noun phrase or verb phrase in its underlying syntactical structure includes a morpheme that ascribes a presentness-including property.

In the 1940s and 1950s, the thesis of linguistic universalism would have been rejected out of hand. The linguistic relativism or "localism" of Sapir, Bloomfield, Whorf, Quine and others was based on *taxonomic grammar*, which concentrated on the surface structure of sentences. But beginning with his 1957 book on *Syntactic Structures*, Chomsky demonstrated that a deep structure needs to be postulated as well. Surface structures vary from language to language, but linguistic universals can be found on the level of deep structure and among the transformational rules that "transform" the deep structure into the surface syntactico-phonological realization of the sentence. Chomsky hypothesized that all humans have an innate and genetically programmed knowledge of linguistic universals and that this is the common element underlying all natural languages. This is a key idea of *transformational grammar,* which superseded taxonomic grammar, and it fits in with the presentist thesis that humans are essentially such that they token only sentences that are about presentness.

But the issue of the exact nature of the deep structure, transformation rules, and the linguistic universals has been a matter of controversy ever since Chomsky first introduced these notions. One difference that became prominent in the late 1960s and early 1970s concerns Chomsky's "standard theory" and Generative Semantics. In *Aspects of the Theory of Syntax* Chomsky supposed that the deep structure of a sentence is a syntactical structure that is different from, and serves as input to, the semantic structure of the sentence. But the Generative Semanticists (McCawley, Lakoff, Harman and others[29]) argued that the deep structure *is* the semantic structure. Generative Semantics has since been superseded by other programs, but one of its key ideas—that deep structure is semantic structure—has been retained by the currently active Metatheory (most fully worked out in William Lycan's 1984 book, *Logical Form in Natural Language*). In the following discussion, I will also assume this identification, although I reject many of the other ideas of Generative Semantics and the Metatheory (since these ideas—for example, the tenseless theory of time, extensionalism, etc.—are inconsistent with the presentist theory developed in the present treatise).

If we assume that deep structure is semantic structure, then the theory of presentism can be placed within the framework of transformational grammar in the following way. In chapter 5 I diagrammed the semantic relations the elements of the A-sentence "John is running" bear toward the constituents of the proposition this sentence expressed (on some occasion of use). We may depart from the many other ways of depicting deep structure by saying that this diagram *is* the depiction of the deep structure of "John is running", relativized to a certain occasion of use. The diagram is Figure 6.1. This diagram is the depiction of the deep structure of "John is running." The deep structure itself is *the proposition* (John as E-related to running and presentness as C-related to John's running) *as articulated into constituents that are relata of the semantic relations of conveyance* (the broken arrow) *and reference* (the unbroken arrows). The transformation rules, which operate upon the deep structure so as to produce the surface structure of the sentence, are rules that op-

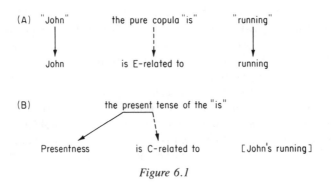

Figure 6.1

erate upon these propositional *relata* so as to produce their surface syntactico-phonological expression by "John is running."

The claim the presentist makes about all natural languages is twofold. The presentist claims that the deep structure of every natural language sentence (as used on some occasion) contains (a) at least one propositional part, (b) at least one propositional E or C relation, (c) at least one semantic reference relation whose *relatum* is the propositional part, and (d) at least one semantic conveyance relation whose *relatum* is the propositional E or C relatedness of the relevant propositional part(s). The second claim is that the deep structure of every sentence contains (e) presentness as a propositional part, (f) a reference relation whose *relatum* is presentness, (g) a C or E propositional relation that has presentness for one of its relational terms, and (h) a conveyance relation whose *relatum* is the propositional E or C relatedness of presentness to something.

In virtually all cases the deep structure of a sentence (as used on some occasion) will contain two or more propositional parts, one of which is presentness. But in sentences that ascribe presentness to presentness, presentness is the only propositional part. Consider that it is false that presentness is passed away, is still to come, or is timeless; it is true that "Presentness is present." In this sentence, "Presentness," "present," and the present tense of the copula refer to the same propositional part—presentness—and the pure copula serves to convey that presentness is E-related to itself (that presentness exemplifies itself). This is the simplest possible proposition, since it is the only proposition that contains only one part.

The presentist thesis about the semantic unity of all natural languages is best stated in terms of completed natural languages, where a completed natural language is an actual natural language plus its extension to cover all propositions. An actual natural language does not have enough words in use to refer to all the infinitely many distinct substances and properties, but it is logically possible to invent enough neologisms so that the language becomes capable of expressing all propositions. For example, several decades ago the English language was increased by adding the name "quark" so that a newly discovered natural kind of substance could be designated. Even if the number of possible neologisms is too small to cover the continuum-many or more substances and properties, the same word can be used in different senses to refer to any number of different substances or properties. Any natural language can be completed in this way and thus any natural language can have its expressive power enhanced to the point where it can express all propositions. Now the presentist thesis about the semantic unity of all languages is a thesis about completed natural languages. The thesis is

(NL) For any pair of completed natural languages Li and Lj, and for any sentence S in Li, there is a sentence S′ in Lj that has the same deep structure as S.

Thesis (NL) entails, given my definition of deep structure, that if a sentence in one completed natural language Li expresses a proposition P, then, for any other natural language Lj, there is a sentence in Lj that expresses P. It might be thought that (NL) is an implausibly strong thesis, but in fact it is weaker than some other theses about natural languages currently endorsed by transformational linguistics. For instance, H. Steklis and S. Harnard[30] hold that

(1) Anything that can be said in one natural language can be translated exactly into any other language.

Steklis and Harnad are referring to actual, not completed, natural languages. Given this, (1) seems clearly false, for some languages contain primitive predicates for which there are no synonyms in another language. For example, the Malagasy predicate "mikabary" refers to the property of performing a *kabary*, which is a particular type of formal speech only given on certain types of ceremonial occasions, such as bone-turning ceremonies.[31] Steklis and Harnad argued that the explanation of the respect in which the foreign sentence is untranslatable is in fact a translation of it, but this is a confusion of a meta-linguistic explication of a sentence in an object language with a synonym in an object language.[32]

A weaker thesis is Jerrold Katz's so-called "determinancy thesis":

(2) For any pair of natural languages Li and Lj, and for any sentence S in Li, and for any sense α of S, there is at least one sentence S' in Lj such that α is a sense of S'.[33]

By a sense Katz means a proposition. Thesis (2) does not entail that every proposition expressed by S (e.g., if S is ambiguous) should be expressed by some sentence S' in another language, but only that one of them be expressed. Keenan has argued that (2) is too strong, since some languages express propositions not expressible in others. Keenan provides several examples, but we may return to the example a sentence in Malagasy which predicates "mikabary" of a certain person P that is directly referred to by a proper name, say "P." This Malagasy sentence will express a *de re* proposition that no sentence in natural English expresses. Keenan proposes instead a much weaker thesis that is relevantly similar to (NL), namely,

(3) Any language can provide exact translations of anything in any other language if we allow ourselves to increase its expressive power.[34]

I think (3) is correct, but I would point out that Keenan's criticism of Katz's thesis (2) is based on a misunderstanding of Katz's theory, since (2) is not about actual natural languages but completed languages. Thus, Katz writes that so-called failures of translation, such as of attempts to translate nuclear physics into a "jungle language," are really "vocabulary gaps" and not failures of translation in Katz's sense:

the failure represents a temporary vocabulary gap (rather than a deficiency of the language) which makes it necessary to resort to paraphrase, creation of technical vocabulary, metaphorical extension, etc., in order to make translations actual in practice, as well as possible in principle.[35]

This suggests that Katz has in mind some principle of translation such as Keenan's (3) and our principle (NL).

But the issue arises of exactly what is meant by "translation" as this is applied to different natural languages. I think (3) and Katz's principle are true if we define "translates"—as this expression is applied to sentences of different languages—as meaning "has the same deep structure as," where "deep structure" is defined in the manner I suggested. Note that "translates" and "is synonymous with" as used to refer to an inter-language relation has a different meaning than it has when used to refer to an intra-language relation. So far in this work I have been concerned with English only and so have used the intra-language definition (where "translates" entails, among other things, a syntactico-semantic correlation). The notion of syntactico-semantic correlation is the main stumbling block for the application of the intra-language definition of translation to inter-language relations, for this notion presupposes *a similarity of surface structure* between sentences that is found in the same language but not between some natural languages. Admittedly, there are syntactico-semantic correlations among some natural languages having the same origin, such as among some Indo-European languages, but this is not true for other natural languages. It is worthwhile to introduce some examples here. Let us first consider some examples that preserve syntactico-semantic correlation.

The sempiternally tensed English sentence "Snow is white," where the "is" means "has always been, is and will always be," is syntactico-semantically correlated with the French and German sentences

(1) La neige est blanche.

(2) Der Schne ist weiss.

The noun phrases "snow," "la neige," and "der Schne" all refer to the property of being snow; "white," "blanche," and "weiss" all refer to the property of whiteness; and "is," "est," and "ist" convey the propositional C-relatedness of being snow and being white and ascribe sempiternality to snow's being white.

Another example of correlation among Indo-European languages concerns the future tensed sentence "I shall speak," which is synonymous and correlated with (for example) the Spanish and Italian sentences

(3) Yo hablare

(4) Io parlero

The future tensed grammatical forms of "hablere" and "parlero" have the same semantic content as the English auxiliary verb "shall"; they ascribe present futurity to my speaking.

Analogously, the present tensed "I have" syntactico-semantically correlates with these Swedish, Dutch and Icelandic sentences

(5) Jag har

(6) Ik heb

(7) Ek hefi

The present tensed grammatical forms of "have," "har," "heb," and "hefi" all ascribe presentness to my having.

There are even syntactico-semantic correlations among some Indo-European sentences and some non-Indo-European sentences, such as Eskimo sentences. In Eskimo the copulae and verbs are not grammatically tensed, but instead the endings of substantives are altered to indicate the temporal location of the matter reported. An example in Eskimo concerns the word "Puyok," meaning smoke. Otto Jespersen points out in his *Philosophy of Grammar* that in Eskimo, from "puyok 'smoke' is formed a preterit puyuthluk 'what has been smoke,' and a future puyoqkak 'what will become smoke'."[36] Such differences do not preclude synonymy and correlation. The natural English sentence "The fire has been smoking" is synonymous with the Eskimo sentence that includes in the relevant way a word for fire and the word for what has been smoke, "puyuthluk." There is a syntactico-semantic correlation, for the grammatical tense of the copula "had been" in the English sentence is correlated with the preterit form of the noun "puyuthluk" in the Eskimo sentence. Both sentences express the same proposition. If the sentences are both used *de re,* they will both express the proposition consisting of the fire as E-related to the property of smoking and of pastness has C-related to the fire's smoking.

However, syntactico-semantic correlation breaks down for many synonymous sentences in different languages, even if they are both Indo-European. Consider these three Russian sentences paired with their English translations:

(8) On bil offitser
He was an officer

(9) On boodyet offitserom
He will be an officer

(10) On offitser
He is an officer

In these three cases, "On" correlates with "He" and "offitser/offitserom" with "an officer." But although there is a correlation of copulae in (8) and (9), there is not in (10). "Bil" correlates with "was" and "boodyet" with "will be," but there is no grammatically present tensed copula or verb in the Russian sentence (10) that correlates with "is." In this case, the *absence* of a copula or verb in the Russian sentence "On offitser" serves to express the ascription of the A-property of presentness. But unless we are to artificially extend the notion of syntactico-semantic correlation we cannot say that this absence is a part of the sentence "On offitser," a part that correlates to the English "is." The more plausible position is that the Russian and English sentences in (10) have the same deep structure but do not (in respect to their surface structure) syntactico-semantically correlate. The deep structure of these two sentences, relativized to an occasion of use, consists of the *de re* proposition consisting (say) of John as E-related to *being an officer* and presentness as C-related to John's being an officer. It is also consists of the reference and conveyance relations of which John, *being an officer,* presentness and the C and E relatedness are *relata.* The main difference is that English has transformational rules that generate the surface locution "*is*" as the relational term of the reference and conveyance relations whose other terms are presentness and the E and C propositional relatedness respectively, whereas the Russian transformational rules generate the surface phenomenon consisting of the *immediate juxtaposition of the pronoun and indefinite noun phrase* that serves as the relational term of these reference and conveyance relations.

It may be obvious that "He is an officer" translates "*On offitser*" but it is important to formulate criteria for translation that can be used for less obvious cases (I will shortly consider a less obvious case from Hopi). It will be recalled from chapter 1, that there are four criteria of translation: sameness of confirmation conditions, truth conditions, logical equivalence, and logical identity. These conditions also apply to inter-language translation, except that the metacriteria for logical identity are different. As explained in the last chapter, the metacriteria for logical identity (i.e., the criteria for deciding when two sentences meet the criterion of being logically identical) in intra-language translations are *intersubstitutivity in belief-contexts* and *syntactico-semantic correlation*. For inter-language translations, we retain intersubstitutivity (as applied to beliefs of bilingual speakers) but use the criterion of *simplicity* in cases where the correlation requirement does not apply. Regarding the Russian sentence "On offitser," we use as our translation the simplest English sentence that has intersubstitutivity in belief-contexts, logical equivalence, and confirmation and truth condition sameness. Since "He is an officer" is simpler than "He is human and an officer" or "He is self-identical and an officer," it alone serves as the translation. But simplicity must be used only in a way compatible with our knowledge of other aspects of the languages. For example, the English "He is an officer" might be the simplest translation of the Russian for "He is human and an officer," but it is not consistent with our knowledge that the former sentence fails to include a synonym for the Russian word for "human."

It is worthwhile to consider how a presentist would meet a specific challenge to his thesis about deep structure, his thesis being that *the deep structure of every sentence (as used on any occasion) in every possible language includes presentness as a part*. Stuart Chase, a linguistic relativist of the Whorf school, claims that some sentences of the Hopi language *ascribe no temporal properties at all* and in fact *have no logical subject at all*. They merely refer to something, without ascribing it any property (it will be recalled that something is a logical subject only if it is referred to *and* ascribed a property by a sentence). Chase claims the Hopi word "Reh-pi" refers to a flash (e.g., a lightning flash); this word is a sentence in Hopi, such that we have "one word for the whole performance, no subject, no predicate, no time element."[37]

First, I would note that the absence of a subject-predicate distinction must be a merely surface phenomenon of the Hopi sentence and cannot reflect a corresponding simplicity of its deep structure. If the deep structure were simple, "Reh-pi" would not express a proposition and thereby would not be a complete sentence. "Reh-pi" must have a complex deep structure, such as the structure consisting of the property *being a flash* as E or C related to some other item. The property of *being a flash* as propositionally related to this other item could be expressed by an English sentence that has a subject and predicate.

Second, I would note that this deep structure must include, in addition to the property of *being a flash,* some presentness-including property. Given the assumption that "Reh-pi" is a possibly true sentence, we can know a priori that *some* presentness-involving belongs to this deep structure, although which one belongs to it is knowable only through an empirical investigation of Hopi speakers. We can rule out Chase's implied claim that "Reh-pi" is possibly true and ascribes no temporal property by the following argument. If "Reh-pi" is uttered with truth in some possible conditions, the conditions must be one of the following, namely, when and only when a flash

1. is present
2. is past
3. is future
4. is present or past
5. is present or future
6. is past or future
7. is present or past or future
8. is present and was present
9. is present and will be present
10. is present, was present, and will be present

This list could be extended by adding more precise temporal determinations, such as "is present and will be present tomorrow" or "is present, was always present and will always be present." But given the relatively imprecise level of specification involved in 1–10, we can know a priori that if the truth conditions of "Reh-pi" are not any one of 1–10, then "Reh-pi" is not possibly true. We can know, furthermore, that a person bilingual in Hopi and English would believe "Reh-pi" if and only if he believed some corresponding English sentence with the same truth conditions. Suppose (1) is the truth condition of "Reh-pi." Then somebody will assent to "Reh-pi" if and only if she is disposed to assent to "There is presently a flash." This or a synonymous English sentence (e.g., "A flash is occurring") is the simplest English sentence that is intersubstitutable in belief contexts with "Reh-pi"; it is simpler, for example, than "There is presently a flash that is self-identical." Accordingly, "There is presently a flash" would meet the criteria for translating "Reh-pi," namely, having the same truth conditions, observational conditions, logical equivalence, intersubstitutivity in belief-contexts, and simplicity. From these and related facts, we could infer that "Reh-pi" has the same deep structure as "There is presently a flash."

But what *are* the empirical facts about "Reh-pi"? The facts about Hopi show that Chase is mistaken about "Reh-pi," for these facts show that "Reh-pi" does contain a time element and is true if and only if a flash *is present or past.* In Benjamin Whorf's paper on "The Punctual and Segmantative Aspects of Verbs in Hopi," he writes that "Hopi also has three tenses: factual or present-past, future, and generalized or usitive."[38] "Reh-pi" is in the present-past tense (the disjunctive tense that ascribes pastness or presentness). The Hopi have no single past tense and no single present tense but instead a disjunctive past-or-present tense, and they rely on the circumstances of the utterance for the information about whether the event in question is past or present.

Now in natural English there is no copula or verb that is a past-or-present disjunctive tense. But there is in natural English a sentence synonymous with "Reh-pi," namely, "There was or is a flash." Furthermore, we can extend English by stipulating that "is" in some uses has a certain species of disjunctive tense not found in natural English, the present-or-past species, and introduce "There is a flash" as the translation of "Reh-pi." The deep structure of both sentences consists of *being a flash* as C-related to *being present or past,* such that the proposition expressed is true if and only if *being a flash* is co-exemplified with pastness or presentness. The deep structure also consists of the semantic conveyance and reference relations of which these propositional constituents are *relata.* The difference is in the transformational rules that generate the surface syntactical phenomena that are relational terms of these relations.

The above examples illustrate the presentist claim that for each sentence S in completed English, and for each completed foreign language L, there is a sentence S' in L that translates S. This entails that each sentence in each completed foreign language has presentness for a logical subject. This confirms the thesis of presentism, that every sentence has presentness for a logical subject and every state of affairs has presentness for a metaphysical subject. The reference to presentness constitutes the semantic unity of all natural languages and the inclusion of presentness as a part constitutes the metaphysical unity of all states of affairs.

Notes

1. A. N. Prior, *Past, Present, and Future* (Oxford: Clarendon, 1967) Roderick Chisholm, *Person and Object* (London: George Allen & Union).

2. J.J.C. Smart, "Time and Becoming," in *Time and Cause*, ed. Peter van Inweigen (Boston: Reidel, 1980), 3–15; David Lewis, *Philosophical Papers* (New York: Oxford, 1983).

3. D. H. Mellor, *Real Time* (Cambridge: Cambridge University Press, 1981).

4. I argue this in "Personal Identity and Time," *Philosophia* (forthcoming). For an interesting discussion of this issue, see L. Nathan Oaklander, "Delmas Lewis on Persons and Responsibility," *Philosophy Research Archives* 13 (1987–88): 181–87.

5. For a definition of the concrete world whole, see my *Felt Meanings of the World: A Metaphysics of Feeling* (West Lafayette, Ind.: Purdue University Press), chaps. 5–6. Note that the concrete world-whole, the universe, is not a "world" in the sense of possible world theories. A possible world, be it the actual world or a merely possible world, is an abstract object, specifically, a proposition. See my "Tensed States of Affairs and Possible Worlds," *Grazer Philosophische Studien* 31 (1988): 225–35.

6. Hector-Neri Castañeda, "Perception, Belief, and the Structure of Physical Objects and Consciousness," *Synthese* 35(1977): 299.

7. *Webster's New World Dictionary*, 2nd college edition (World Publishing Co., 1974), 1466.

8. Cf. D. M. Armstrong, *A Theory of Universals,* vol. 2 (Cambridge: Cambridge University Press, 1978), 19–23.

9. Cf. Jerrold J. Katz, *Cogitations* (New York: Oxford University Press, 1988).

10. Oaklander actually says they are not "timeless" but "eternal," since they do not obtain *in* time but involve temporal relations. See his *Temporal Relations and Temporal Becoming,* 19.

11. D. M. Armstrong, *What Is a Law of Nature?* (Cambridge: Cambridge University Press, 1983); Michael Tooley, "The Nature of Laws," *Canadian Journal of Philosophy* 7(1977): 667–98; Fred Dretske, "Laws of Nature," *Philosophy of Science* 44(1977): 248–68.

12. Armstrong, *What Is a Law of Nature?*, 78.

13. Alfred J. Freddoso, *The Necessity of Nature,* Midwest Studies in Philosophy, no. 11 (Minneapolis: University of Minnesota Press, 1986), 215–42.

14. P. F. Strawson, *Introduction to Logical Theory* (New York: Methuen, 1952), 150.

15. Panayot Butchvarov, *The Concept of Knowledge* (Evanston, IL: Northwestern University Press, 1970), 151.

16. J.J.C. Smart, *Philosophy and Scientific Realism* (New York: Humanities, 1963), 133.

17. Cf. H. P. Grice, "Logic and Conversation," in *The Philosophy of Language,* ed. A. P. Martinich (New York: Oxford University Press, 1985), 159–70.

18. W. Kneale, "Time and Eternity in Theology," *Proceedings of the Aristotelian Society* 1960–61: 98.

19. Butchvarov, *Concept of Knowledge,* 149.

20. Ibid., 148.

21. Gottliebe Frege, "The Thought," in *Essays on Frege,* ed. E. D. Klempke (Urbana: University of Illinois Press, 1968), 534.

22. Nicholas Wolterstorff, "God Everlasting," in *God and the Good,* ed. C. Orlebeke and L. Smedes (Grand Rapids, Mich.; Eerdmans, 1975), 186.

23. Ibid., 186–87.

24. Delmas Lewis, "Eternity Again: A Reply to Stump and Kretzmann," *International Journal for the Philosophy of Religion* 15(1984): 73–79.

25. These criteria are explained and justified my *Felt Meanings of the World: A Metaphysics of Feeling* (West Lafayette, Ind.: Purdue University Press, 1986), 20.2.

26. Cf. D. M. Armstrong, *Universals and Scientific Realism,* vol. 2 (Cambridge: Cambridge University Press 1978), 71–74; George Bealer, *Quality and Concept* (Oxford: Clarendon, 1982).

27. See "A New Typology of Temporal and Atemporal Permanence," *Noûs* 23(1989): 307–30.

28. Noam Chomsky, *Aspects of the Theory of Syntax* (Cambridge: Massachusetts Institute of Technology Press, 1965), 28.

29. J. McCawley, "Meaning and the Description of Languages," in *Readings in the Philosophy of Language* ed. J. F. Rosenberg and C. Travis (Englewood Cliffs, NJ: Prentice-Hall, 1971); G. Lakoff, "On Generative Semantics," in D. Steinberg and L. Jacobovits, *Semantics: An Interdisciplinary Reader* (Cambridge: Cambridge University Press, 1971); G. Harman, "Deep Structure as Logical Form" and "Logical Form," in *Semantics of Natural Language* (Dordrecht: Reidel, 1972).

30. H. Steklis and S. Harnad, "From Hand to Mouth: Some Critical Stages in the Evolution of Language," *Annals of New York Academy of Sciences* (1976).

31. See E. Keenan, "Some Logical Problems in Translation," in *Meaning and Translation,* ed. F. Guenthner and M. Guenthner-Reuttner (London: Duckworth, 1978) 164.

32. Keenan makes this criticism in ibid., 158.

33. J. J. Katz, "Effability and Translation", in *Meaning and Translation,* op. cit., 206.

34. Keenan, "Some Logical Problems," 174.

35. Katz, "Effability and Translation," 220.

36. Otto Jesperson, *The Philosophy of Grammar*, 283.

37. Stuart Chase, Foreward to *Language, Thought and Reality* ed. John Carroll, (Cambridge, MA: MIT Press, 1956), p. viii.

38. Benjamin Whorf, "The Punctual and Segmentative Aspects of Verbs in Hopi," in *Language, Thought and Reality*, 51.

7

Absolute Presentness and the Special Theory of Relativity

7.1 The Special Theory of Relativity

Presentism implies that every state of affairs is of a form such as *Presentness-inheres-in-such-and-such* or *Presentness-inheres-and-always-will-inhere-in-such-and-such*. This implies that states of affairs are not of such forms as Presentness-inheres-in-such-and-such-relativity-to-a-reference-frame-R or Presentness-inheres-and-always-will-inhere-in-such-and-such-relativity-to-a-reference-frame-R. In other words, presentism entails that presentness inheres in events *absolutely,* not relatively to a reference frame. The presentness of events is an absolute presentness. Using the metaphor of a "tide of becoming," we may say that presentism is the theory that there is a single, universewide "absolute tide of becoming," instead of many criss-crossing tides of becoming, one for each reference frame.

The preceding remarks suggest that presentism is true only if Einstein's Special Theory of Relativity (STR) is false. But this suggestion needs to be made more precise. For one thing, the STR (as originally formulated by Einstein in his 1905 paper and as formulated in a space–time framework by Minkowski in his 1908 paper) is constructed from the philosophical perspective of the tenseless theory of time. Accordingly, it is not obvious that and how the STR entails that the A-property of presentness cannot inhere in events absolutely. Second, it is not clear in what sense a presentist could reasonably argue that the STR is false, given that the STR has been extensively confirmed by observations. These two issues are the focus of this section. Following this, the presentist argument for the absoluteness of presentness will be constructed.

Although Einstein and Minkowski formulated the STR from the framework of a tenseless theory of time, the physical content of the STR can be abstracted from this framework and placed in the context of the tensed theory of time. This has been done in various ways by Howard Stein, E. Harris, Storrs McCall, Paul Fitzgerald, William Godfrey-Smith, Ferrel Christensen, William Lane Craig, Lawrence Sklar, D. Dieks, and others.[1] The problem that the STR poses to presentism can be exactly stated once we reformulate the STR in tensed terms. I shall give three different tensed formulations of the STR and show how, in each case, we are forced, on pain of contradiction, to conclude that presentness inheres in events relatively to a reference frame and therefore that presentism is false.

The three different interpretations can be explained in terms of a diagram of the light cone of the STR (see Figure 7.1). Suppose that E is an event occurring here and now. The lower cone in which F is situated is the "absolute past" of E. This means that event F and every other event in this cone is such that a physical influence emanating from the event is able to influence the event here and now, event E. The cone is called a "light cone" since no physical influences are able to exceed the speed of light; consequently, the cone includes all—and only—the events connect-able with E by physical influences that travel at, or slower than, the speed of light. Event D and all other events in the upper light cone belong to the "absolute future" of E; they include all—and only—the events that may be influenced by physical processes emanating from E.

The three tensed interpretations of the STR differ in respect of their views about the events in the "absolute elsewhere" of event E. Events in E's absolute elsewhere include events A, B, C, and G and all other events that fall outside of E's light cone. No events that fall outside of E's light cone can either affect, or be affected by, E. This implies that no light signal sent from A, B, C, or G can reach event E and that no light signal sent from E can reach these events. Events A, B, C, and G, consequently, are neither in E's absolute past nor in E's absolute future. Are they present simultaneously with E?

A. All Elsewhere Events Are Present

The first interpretation I shall consider holds that events A, B, and C and all other events in E's absolute elsewhere are simultaneous with E. If E is present, then A, B,

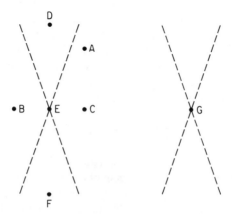

Figure 7.1   The letters represent events, and the slashed lines are the boundaries of the light cones of events E and G.

C, and G are present. This interpretation of the STR is discussed by Robert Weingard and is mentioned by Ferrel Christensen, D. H. Mellor, Nicholas Maxwell, and Paul Fitzgerald.[2] It can easily be seen that this tensed interpretation entails (on pain of contradiction) that presentness inheres in events only relatively to a reference frame. Assume for reductio that presentness inheres in events absolutely. Suppose that E is present. It follows that D is future, since D is in E's absolute future. But if E is present, G is also present, since G falls outside of E's light cone. However, if G is present, then D is present, for D falls outside of G's light cone, which contradicts the earlier statement that D is future. This contradiction disappears if the inherence of presentness is relativized to different reference frames. Consider the reference frame occupied by a possible observer who is first at event F, then at E, and later at D. Call this reference frame R1. Call R2 a reference frame occupied by a possible observer who is first at an event in G's past light cone, then at G, and later at an event in G's future light cone. The inherence of presentness may then be relativized to R1 and R2, respectively. The following temporal descriptions may then be introduced:

(1) If E is present relatively to R1, then D is future relatively to R1.

(2) If E is present relatively to R1, then G is present relatively to R1.

(3) If G is present relatively to R2, then E is present relatively to R2.

(4) If G is present relatively to R2, then D is present relatively to R2.

Descriptions (1) and (4) are mutually consistent, since it is possible for D to be both future *relatively to R1* and present *relatively to R2*, given that E is present relatively to both R1 and R2.

B. All Events Orthogonal to the Time Axis of a Present Event Are Present

The second interpretation holds that only some of the events that are elsewhere from E are present if E is present. Consider the spacelike hypersurface x orthogonal to a time axis t, such that t is the time axis that runs through F, E, and D (see Figure 7.2). The hypersurface x is the line that runs through B, E, C, and G. The second tensed interpretation holds that if E is present, then all and only the events on this hypersurface are present. (This interpretation is developed in different ways by Storrs McCall and Lawrence Sklar and is also discussed by Paul Fitzgerald.)[3] This implies that if E is present, then B, C, and G are present but D is not present. Thus, the aforementioned contradiction is avoided.

But a new contradiction takes its place (if presentness is assumed to be absolute). Consider a time axis t' at angles to t, such that the spacelike hypersurface orthogonal to t' at E is x' (see Figure 7.2). If E is present, then A is present, since A is located on a hypersurface orthogonal to the time axis t' on which E is located. This entails a contradiction, since A is also future, considering that it is above the hypersurface x, which is orthogonal to the time axis t at E. This contradiction is resolved if we suppose that presentness and futurity inhere in events relatively to a

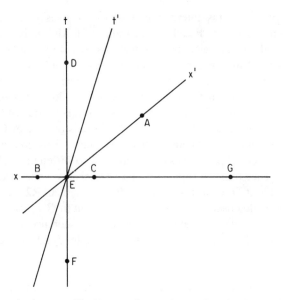

Figure 7.2   x is the spacelike hypersurface orthogonal to the time axis t. x' is the
              spacelike hypersurface orthogonal to the time axis t'.

reference frame. If R1 is the reference frame whose time axis is t and R3 is the
frame whose time axis is t', then we can consistently say that if E is present
relatively to both R1 and R3, then A is future relatively to R1 but present relatively
to R3.

C. *Only What Is Here-Now Is Present*

The third interpretation is that if E is present, then only E is present. Events in the
future light cone of E are future, and those in the past are past; but all events in E's
absolute elsewhere have no A-position at all. They are A-indeterminate. This interpretation is advocated by Howard Stein, E. Harris, and D. Dieks and is mentioned
by Nicholas Maxwell and D. H. Mellor.[4] If E is present, then events A, B, C, and G
are neither present, nor future, nor past. The reasoning behind this third interpretation is that A, B, C, and G are not connectable with E by any physical process and,
consequently, that their temporal relations to E are, in principle, unverifiable by an
observer located at E, who thus must regard their temporal positions as indeterminate
relative to E. This does not mean that the observer O at E merely does not know which
temporal relations obtain between E and the events in E's absolute elsewhere but that
relatively to O at E, there are no temporal relations between E and these events.
Temporal relations obtain only at later times, when events, A, B, C, and G become
parts of O's absolute past and are connectable with O via physical influences.

    This third interpretation enables one to avoid the contradictions entailed by the
first two interpretations (understood absolutely); for if E is present, then no ascriptions of A-properties to A, B, C, D, and G are possible and therefore no ascriptions
of incompatible A-properties to these events are possible.

Nevertheless, a new contradiction arises on the third interpretation if A-positions are held to be absolute. Suppose that an observer O occupying the reference frame R1 is at E and an observer P occupying the reference frame R2 is at G. According to the observer O, E is present and G is temporally indeterminate. But according to observer P, G is present and E is temporally indeterminate. Since being temporally indeterminate implies not being present, two contradictions emerge if these A-descriptions are interpreted absolutely: "E is both present and not present" and "G is both present and not present." To avoid the contradictions, a relativization of the temporal positions is required, so that we have, instead, the descriptions "E is present relatively to O at E and not present relatively to P at G" and "G is present relatively to P at G and not present relatively to O at E."

If the STR and the tensed theory of time are both true, then one of these three tensed interpretations of the STR must be the true theory of time.[5] But if one of these interpretations is the true theory of time, then presentism is false, since presentism holds that A-properties are possessed absolutely. Since the presentist is committed to the tensed theory of time, the only course open to him is to argue that the STR is false.

It might be thought to be a hopeless task to show that the STR is false. Is not the STR one of the best-confirmed physical theories of this century? Bondi notes that "there is perhaps no other part of physics that has been checked and tested and cross-checked quite as much as the Theory of Relativity."[6] Among the evidence for the STR is the Thomas precession, the Kennedy–Thorndike experiment, the ionization of charged particles, the Ives–Stilwell experiments, the observed lifetime of muons, the Mossbauer effect and the Hafele–Keating experiment.[7] (Of course, there is the real or alleged threat posed to the STR by Aspect's confirmation of Bell's theorem, but we may ignore this complication for present purposes.) However, there is a line of argument open to the presentist that avoids the obvious problem that "there is empirical evidence that supports the STR." It could be argued that the STR is false inasmuch as it purports to be a theory about time but that there is a true theory—call it the Nontemporal STR—that is just like the STR except that it purports to be a theory about the observable behavior of light rays, rigid bodies, and the like but does not purport to be a theory about time. It could be argued that the empirical evidence, in fact, supports not the STR but the Nontemporal STR. Light rays, rigid bodies, and the like are observed to behave in the manner described by the STR; but (contra the STR) this does not entail anything about the nature of *time*.

This is the line of argument I shall develop in the remainder of this chapter. I shall argue that the STR is false and that the Nontemporal STR is true. In the next section I shall outline the strategy of this argument.

7.2 Metaphysical Time and Special-Theory-of-Relativity-Time

My strategy is to argue that what the STR calls "time" is, in fact, not time but a nontemporal system and that *time* is absolute, rather than relative. It will be recalled that in chapter 6 I stated that "exists in time" may be used in wider or narrower senses. These senses correlate with the different degrees of completeness of the description of something. The maximal degree of completeness is the description of all the states of the object, including all its relational states. Something "exists in time" in the broadest sense if temporal predicates are required to describe any of the object's states, including such relational states as the exemplifications of relational properties of *being referred to*. Let us call this broadest sense of time *metaphysical time*. Something exists in metaphysical time if and only if it acquires or loses any n-adic property. I argued in chapter 6 that propositions exist in metaphysical time on the ground that at some times they exemplify *being referred to* and at later times do not. Narrower senses of "exists in time" correlate to less complete descriptions of an object, such as a description that mentions only states involving real (rather than Cambridge) changes in an object or only states involving luminal or subluminal physical relations with other objects. One of these narrower senses of "exists in time" is the sense that "time" has in the STR, and time in this sense may be called *STR time*.

I would now like to supplement and qualify the ideas of chapter 6 by pointing out that if three conditions are met, it may be soundly argued that *only one* of the senses of "exists in time" is a truly *temporal* concept and that the other senses are really nontemporal concepts. For example, if metaphysical time meets the following three conditions, then metaphysical time may be identified with *time,* and the other senses of "exists in time" may be taken as nontemporal concepts:

1. The events that are temporally related to each other in metaphysical time are (qua related in metaphysical time) related by *primitive* temporal relations, such that "earlier," "later," and "simultaneous" must be taken as indefinables in referring to these relations. The same holds for the A-properties.
2. The events that are related to each other in the other alleged "times" are (qua related in these "times") related by *reducible* temporal relations, such that "earlier," "later," and "simultaneous" as referring to these relations can be defined in terms that contain no temporal words (except perhaps for words referring to the primitive temporal relations of metaphysical time). The same holds for the A-properties.
3. All the events that are related by the alleged "temporal relations" of the other alleged "times" are also related by the primitive temporal relations of metaphysical time. The same holds for the A-properties. In other words, metaphysical time is *global*, or all-embracing.

An example will help clarify these conditions. If the event E1 is "earlier than" E2 in the STR sense that E1 can causally influence E2 and also *earlier than* E2 in the primitive temporal sense of metaphysical time, then the STR causal relation is not a temporal relation but a *causal relation,* and the only temporal relation between E1 and E2 is the primitive temporal relation of *earlier than* that belongs to metaphysical time. The claim that the STR causal relation is a temporal relation of *earlier than* may be true if there is no primitive relation of *earlier than;* but if there is such a primitive relation and conditions 2 and 3 are also met, then it is false that the STR causal relation is a *temporal* relation.

If there is no sense of "exists in time" that satisfies all three of the aforementioned conditions, then it cannot be soundly argued that there is only one proper sense of "exists in time." Consider the three conditions mentioned:

1. If the temporal predicates "earlier," "past," and so on did not have a primitive temporal meaning for any of the senses of "exists in time," then there would be no basis for singling out one of the senses as the concept of time and the other senses as concepts of something else.
2. If there were two or more senses of "exists in time" that involved primitive meanings of these temporal predicates, then there would be no basis for deciding which of these senses was "the" temporal sense.
3. If there were only one sense of "exists in time" that involved primitive temporal predicates but some events that obviously exist in time do not exist in time in this sense, then we could not say that the primitive sense is *the* concept of time. For example, if propositions exist in time in a primitive metaphysical sense but physical events exist in time not in this primitive sense but only in the STR sense, then it would be false that metaphysical time and only metaphysical time is *time*.

I think it can be shown that metaphysical time is the only primitive time and that the temporal adjectives ("earlier," "past," etc.) used in theories of "time" in all its other senses are definable in nontemporal terms or with a conjunction of nontemporal terms and the primitive temporal terms of metaphysical time. Furthermore, I think that anything that "exists in time" in any of the other senses also exists in time in the metaphysical sense. If this can be shown, it will follow that metaphysical time *is* time and that other senses in which "time" can be used are, in fact, nontemporal concepts, such as causal concepts, concepts of change, concepts of psychological experiences, and concepts of light-connectibility. I shall argue for these claims in the following sections, taking as my prime examples of reducible time STR time and psychological time.

Accordingly, my argument that time is absolute and that STR is false has the following meaning. My argument is that STR time is not, in fact, time and thus that the STR shows not that time is relative but merely that certain light-connectibility relations are relative. I shall argue that *time* (i.e., time in the primitive sense, *metaphysical time*) is absolute. I begin by showing that the STR temporal n-adic properties are reductively definable in terms of nontemporal properties and some primitive temporal properties of metaphysical time (section 7.3). Following this, I show that the temporal properties of metaphysical time are both primitive and absolute (section 7.4). It is then argued that all mental events exist in metaphysical time and that psychological time is reducible to psychological properties and the primitive temporal properties of metaphysical time (section 7.5). The argument is then advanced that all physical events exist in metaphysical time (section 7.6). I then argue that metaphysical time is the only time in each possible world in which there is time (section 7.7). I conclude the discussion by distinguishing metaphysical time from other alleged candidates for "absolute time," such as Newtonian time (section 7.8). If my arguments are successful, I will have established one of the main theses of presentism, namely, that presentness inheres in states absolutely, rather than relatively.

7.3 The Reductionist Nature of the Special Theory of Relativity Temporal Definitions

It suffices for my purposes to consider only the definitions of "simultaneity" that are belong to the STR. There is not one such definition but three: topological simultaneity, distant simultaneity, and local simultaneity. I shall show that these definitions are reductionist in the sense that they define "simultaneity" in terms of nontemporal words or in terms of a conjunction of nontemporal words and words that refer to a primitive temporal relation that belongs to metaphysical time; that is, I shall show that STR time reduces to something nontemporal or to a conjunction of something nontemporal and metaphysical time.

For the sake of illustrative clarity, I use the letters A–G in the three definitions to depict the events represented in Figures 7.1 and 7.2. Topological simultaneity is defined as follows:

> D1 E and A are topologically simultaneous = df. E and A are unconnectible by luminal or subluminal signals.

D1 is clearly reductionist, for it defines temporal words in terms of nontemporal words. The sentence "E and A are simultaneous" (in the topological sense) uses "simultaneous" not as primitive but to express the concept of luminal/subluminal unconnectibility, such that the sentence is synonymous with "E and A are unconnectible by luminal or subluminal signals."

The STR definition of distant simultaneity is

> D2 B and C are distantly simultaneous = df. B and C emit (or might have emitted) light signals that arrive simultaneously at the observationally verifiable midpoint D between B and C, relatively to a reference frame R1.

I think it is clear that D2 is a reductionist definition, since its *defiens* contains only nontemporal words except for "simultaneously" and since "simultaneously" refers to local simultaneity, which is reductively definable (as I shall now show).

Locally simultaneous events are not represented in my diagrams; but we may suppose there is some event E' arbitrarily close to E and that is locally simultaneous with E in this sense:

> D3 E and E' are locally simultaneous = df. E and E' occur approximately at the same place and approximately at the same time.

Local simultaneity is partly definable in terms of the nontemporal property of being approximately at the same place, but D3 will not be reductionist in my sense unless "approximately at the same time" also refers to a nontemporal relation or else refers to a primitive temporal relation of metaphysical time. It is clear that a part of this expression refers to some nontemporal relation, namely, *approximation,* and that the crucial matter concerns the referent of "at the same time." This phrase cannot refer to distant simultaneity or topological simultaneity, since the latter two relations only obtain between events at different places. Nor does it refer to any sort of relative simultaneity, because if it two events occur at the same place and time, this is an absolute matter. It seems to me that in the statement that two events occur at the same place and at the same time, "at the same time" refers to a primitive and dyadic simultaneity relation. This is the very relation of absolute simultaneity that belongs to metaphysical time, as I shall argue in the next section.

Reichenbach endeavors to analyze local simultaneity in terms of the relation of identity. According to him, "at the same time" refers to a relation of *identity* that obtains between time positions, not to a relation of simultaneity. He writes:

> We shall distinguish between the *simultaneity at the same place* and the *simultaneity of spatially separated events*. Only the latter contains the actual problem of simultaneity; the first is strictly speaking not a simultaneity of time points, but an *identity*. Such a concurrence of events at the same place and at the same time is called a *coincidence*. In a strict coincidence there is actually no comparison of space or time since position and time are identical for both events. Practically speaking, such an identity never occurs since we could no longer distinguish the two events. But an approximate coincidence can be realized, in the example of two colliding spheres or two intersecting light rays.)[8]

We may agree with Reichenbach that in strict coincidence, events are not "simultaneous" if this word refers to a relation between events at different space–time positions, that is, if it refers to distant or topological simultaneity. And we may agree that if two events E and E' occur at the same space–time position, that entails "The time position of E is identical with the time position of E'." However, I would add that this latter sentence is logically equivalent to the assertion that E and E' sustain a primitive relation of absolute simultaneity. "Absolutely speaking, the time point at which E occurs is identical with the time point at which E' occurs" is logically equivalent to "Absolutely speaking, E and E' are simultaneous" (where "simultaneity" expresses a primitive and dyadic temporal relation). If strictly coincident events are regarded as identical, so that E = E', then this statement remains true, since absolute simultaneity is an equivalence relation; that is, it is symmetrical, transitive, and reflexive, such that each event or time point has this relation to itself. These considerations suggest that it is reasonable to think that the primitive and absolute relation of simultaneity that belongs to metaphysical time is a part of the reduction basis of local simultaneity; it is the referent of "at the same time" that occurs in the defiens of D3.

In summary, topological simultaneity is reducible to a nontemporal relation, and distant simultaneity and local simultaneity are both reducible to a complex of nontemporal relations and a primitive relation of metaphysical simultaneity. (Since distant simultaneity is defined partly in terms of local simultaneity and since the latter is defined partly in terms of metaphysical simultaneity, distant simultaneity is defined partly in terms of metaphysical simultaneity.)

Analogous remarks could be made about the relations of earlier and later and about presentness, pastness, and futurity as these are defined in one of the three tensed interpretations of the STR. A case in point is a definition of presentness that is implied by the second tensed interpretation of the STR discussed in section 7.1. This is a definition of *distant presentness* and correlates with the definition D2 of distant simultaneity:

D4 B is distantly present relatively to the reference frame R1 = df. Some event E that is located on the time axis of R1 is present relatively to R1, and B is distantly simultaneous with E.

Since "distant simultaneity" in the *defiens* is reductively definable (as just indicated), this makes D4 at least partly reductive. The extra ingredient in the tensed definition is the notion of "is present relatively to R1." This mentions the nontemporal property of *relatively to a reference frame,* and so this ingredient is at least partly reductive. What is the referent of "present"? I think that "present" should be taken as referring to the primitive monadic A-property that is constitutive of metaphysical time. The different between "is present" and "is present relatively to R1" is that the former refers to monadic presentness alone, the latter, to monadic presentness *as exemplified relatively to a reference frame.* If we want to say that "presentness is a dyadic property if the STR is true" (as I said in section 1.5), we must interpret this as meaning that if the STR is true, then the monadic property of presentness is exemplified only relatively to a reference frame, such that it is exemplified only as a part of the complex dyadic property ( ) *being present relatively to* ( ). If some event E exemplifies this complex property, then E exemplifies presentness, and E's exemplification of presentness itself exemplifies *being relative to* ( ).

This suggests that the definitions of the STR A-properties are reductive in the required sense, for they define these properties in terms of a conjunction of nontemporal properties and the primitive temporal properties of metaphysical time. Given that the STR temporal relations and properties are reducible to nontemporal properties and relations and the primitive temporal properties or relations of metaphysical time, one of the conditions of demonstrating the falsity of the STR is met. (Remember that "the falsity of the STR" in the present context means "Time does not have the nature attributed to it by the STR, the true theory, instead, being the Nontemporal STR.") The next tasks are to show that the properties and relations of metaphysical time are (1) primitive, (2) absolute, and (3) such that they are exemplified by all the events that satisfy the STR definitions and any other alleged temporal definitions.

7.4 The Primitiveness and Absoluteness of Metaphysical Time

I want to suggest first that the relations and properties of metaphysical time are primitive. I think it a general principle that we should understood temporal-relation expressions and property-expressions in a primitive sense unless there is reason to think otherwise. If I say "x is simultaneous with y," I am properly understood to be using "simultaneous" in a primitive temporal sense unless I indicate that there is some reason to think that this word cannot have a primitive meaning (e.g., if I am discussing the STR). My suggestion is that there is no reason to think that the temporal-relation expressions and property expressions used in describing metaphysical time are not being used primitively and thus that it is reasonable to believe they are being used primitively. This can be elucidated as follows.

I argued in chapter 6 that necessarily true *de dicto* propositions exist in metaphysical time and are always present. Since "x is present and y is present" entails "x and y are temporally located at the same time" and since the latter entails "x and y are simultaneous," it follows that any two necessarily true *de dicto* propositions are simultaneous. How should we understand "simultaneous" here? Clearly, "simultaneous" cannot mean what it does in the STR, namely, D1–D3, since each of these definitions is satisfied only by spatially located events, and propositions are not spatially located. I suggest that "simultaneous" as used to describe a temporal relation between two propositions should be understood in a primitive sense. "Simultaneous" means *simultaneous* and nothing else. In order to deny this, one must provide some reason why "simultaneous" cannot be understood in a primitive sense in this context. One must provide some reductive definition of "simultaneous" and show that this definition is satisfied by the two propositions and that the proposition's satisfaction of this definition entails they do not bear a primitive relation of simultaneity to each other. It seems to me that this cannot be done, so that we are warranted in believing that they are primitively simultaneous. At best, one may be able to find some relation R that is necessarily coextensive with the simultaneity relation among propositions; but that fact, by itself, would no more show that the simultaneity relation *is* R than "Size is necessarily coextensive with shape" shows that "size *is* shape." However, once we realize (as we shall shortly) that propositions bear a simultaneity relation to other items as well (e.g., mental and physical events), then it will appear that there is no relation R that is necessarily coextensive with this simultaneity relation. Thus, the belief that this simultaneity relation is primitive is warranted.

Similar remarks hold for the n-adic properties of being earlier, being later, presentness, pastness, and futurity that constitute metaphysical time.

The primitive simultaneity among propositions is also an *absolute,* or dyadic, relation; that is, it has the form *( ) simultaneous with ( ),* rather than *( ) simultaneous with ( ) relative to ( ).* If it were relative, it would obtain relatively to some third term. But there is no candidate for such a third term for the simultaneity relation between two propositions. The reason for this is that propositions are nonphysical. According to the STR, the relativity of distant simultaneity is based upon physical emissions from the simultaneous events and the relative motions of the reference bodies that receive these emissions. The distant simultaneity relation is relative to a reference body that stands in certain spatial relations to the events. It follows that if it is logically impossible for there to be physical emissions from two items x and y and logically impossible for there to be reference bodies that are spatially related to x and y, then it is logically impossible for x and y to sustain the relative simultaneity relations described in the STR. Similar considerations suggest that it is also logically impossible for x and y to sustain any other relative simultaneity relations, considering that all such relations involve physical emissions and relative motions. Consequently, if two nonphysical objects are simultaneous, they must be absolutely simultaneous. It follows that propositions sustain absolute temporal relations to each other. If two propositions are present, they are present simultaneously, and this relation is two-termed.

But these remarks are not quite correct; for there is a concept of relative time that is psychological, rather than physical in nature. We may define "x and y are simultaneous relatively to the person P" as meaning "x and y are perceived in the same complex act of perception experienced by P." And "x is later than y relatively to P" may mean "x is perceived by P and y is remembered by P, such that the perception and memory belong to the same complex experience of P." This psychological time is not only relative but also nonprimitive, since the temporal words used to refer to it are defined in terms of psychological words. But the temporal relations among propositions cannot be these or analogous psychological relations (such as "being objects of the same complex propositional attitude experienced by P"), since propositions are present nondependently upon their being objects of psychological acts. This psychological time might be the time of propositions if conceptualism is true; but if Platonism is true (as I maintain), then the primitive and absolute metaphysical time must be the time in which propositions exist. (If theism is true, then the temporality of propositions might be reductively definable in terms of God's psychological time. But apart from difficulties with the assumption that theism is true,[9] this reduction will not work, for reasons similar to the ones I shall discuss shortly regarding the psychological time of human minds.)

These considerations about propositions as construed platonically give reason to think that relations of absolute simultaneity, where simultaneity is taken as a primitive relation, actually obtain. They give reason to think that presentness inheres in propositions absolutely. But this would not suffice to vindicate presentism, for it could be objected that even if propositions are present absolutely, physical (and perhaps mental) events are not. And if physical events are present relatively, then presentism is false; for it would then be false that each present event is present absolutely. These difficulties pertain directly to my criteria for establishing that metaphysical time is the only time and that STR time is not time at all. I said in section 7.2 that if we are to establish that metaphysical time is the only time and that STR time is not time at all, we must establish not merely that STR time is reductive and metaphysical time primitive but also that *everything that exists in STR time also exists in metaphysical time*. Given that the relations and properties of metaphysical time are both primitive and absolute, this entails that it be demonstrated that these same relations and properties be shown to be exemplified by physical events. Furthermore, if it is to be shown that metaphysical time *is* time, it must also be shown that all mental events also exist in metaphysical time and that so-called psychological time is no less reductive that STR time.

In the following two sections I will show that mental and physical events exist in metaphysical time. Since one of my arguments that physical events exist in metaphysical time hinges upon the premise that mental events exist in metaphysical time, I shall begin with mental events.

7.5 All Mental Events Exist in Metaphysical Time

There is an argument based on the absolute temporality of propositions that shows that some mental events are terms of absolute temporal relations. Consider all the mental events that are propositional attitudes. My act of thinking of the proposition *that a triangle is three-sided* is simultaneous with this proposition at the time that this act of thinking occurs. When the thinking ceases, the proposition was, but no longer is, simultaneous with this act of thinking. The act of thinking cannot sustain the relative simultaneity relation defined in D3 ("distant simultaneity") to the proposition, since there is no spatial midpoint between the act of thinking and the proposition. Given that no other sort of relative simultaneity relation can be plausibly posited, it is reasonable to believe that this simultaneity relation is an absolute one, that is, a two-termed relation. Given this, it follows that while the thinking is occurring, the thinking is absolutely simultaneous not only with this proposition but with everything else that is then absolutely simultaneous with this proposition. This means that the act of thinking is absolutely simultaneous with every proposition then existing and is absolutely simultaneous with every other propositional attitude that is then absolutely simultaneous with any of these propositions. By virtue of these relations, mental events that are embodied in bodies very distant from one another sustain absolute temporal relations to each other.

It might be suggested that a relative *psychological* relation of simultaneity is the relation that connects the act of thinking (call it A1) to the proposition *that a triangle is three-sided* (call it p). Given this, there is no need to assume that the metaphysical relation of primitive absolute simultaneity connects A1 to p. The relative psychological relation would be that p and A1 are *simultaneous relative to a second act of thinking A2,* where "are simultaneous" means "are coapprehended by" or "are both accusatives of" the second act of thinking A2. There would be a second act of thinking A2 that apprehends both p and A1; and the coapprehension of p and A1 by A2 *is* the simultaneity of A1 and p, relatively to A2.

But this suggestion will not work. For one thing, there is usually no second act of thinking A2 whose accusative is both A1 and p. Accordingly, there is usually no relative relation *being coapprehended by [A2]* that could serve as the simultaneity relation between p and A1. But even if there were, the coapprehension relation could not be the simultaneity relation between A1 and p; for the act of thinking A2 must itself be simultaneous with the proposition p: given that the proposition p is sempiternally present, A2 must occur simultaneously with it if A2 is to apprehend it. What would constitute this simultaneity relation between p and A2? If it is not the absolute simultaneity relation of metaphysical time, it would have to be some relative psychological relation involving some third act of thinking A3, namely, the coapprehension of A2 and p by A3. But A3 must also be simultaneous with p. Clearly, an infinite regress has begun. This shows that acts of comprehending propositions and propositions are not "relatively simultaneous" in the sense of psychological time—if only for the reason that it is a fact that they are simultaneous and it is a fact that there is no infinite regress of comprehensions of propositions.

This argument shows that all propositional attitudes are in absolute metaphysical time by virtue of being temporally related to propositions. But there is also a logically independent argument that shows that every mental event (be it a propositional attitude or something else, e.g., a sensation) is a term of absolute temporal relations. This argument is based on the fact that mental events are temporally related to each other. My memory of something that happened some time ago is occurring simultaneously with my wish that something different happen in the future. It is impossible for a memory and a wish to have physical emissions that intersect at a spatial midpoint between the memory and the wish. (I am assuming that mental events are not spatially located and thus that physicalism is false). It follows that the memory and the wish are not relatively simultaneous in STR time.

This suggests they are absolutely simultaneous. This argument can be generalized. The mental state of every person and animal is a complex whole that consists of component parts. These parts, since they are spatially unlocated, cannot emit physical processes (e.g., light signals) that intersect at a midpoint between them; therefore, the temporal relations that obtain among these parts are absolute relations. Furthermore, the whole mental state that is composed of these parts sustains absolute temporal relations at least to the whole mental states that precede or succeed it and to which it is connected by relations of possible remembrance or expectation. These considerations support the view that there is no mental item that is not a term of an absolute temporal relation.

It might be thought that the mental events are temporally interrelated in a relative psychological time, so that the fact that "mental events are temporally interrelated" need not be explained in terms of their occupation of a primitive metaphysical time that is absolute. For example, a person P's memory and wish could be simultaneous relatively to P in that they both belong to the same complex experience of P. And P's headache could be earlier than P's backache, relatively to P, in that the headache is remembered by P and the backache perceived by P.

But this suggestion will not work, because psychological time is reducible; that is, it reduces to a conjunction of nontemporal properties and the temporal properties of metaphysical time. Indeed, if anything occupies "psychological time," that entails that it occupies metaphysical time. Consider the phrase "The memory and wish occur in the *same* complex experience of P." What does "same complex experience" mean? Clearly, it cannot mean an experience that has some parts earlier than others; for if it did, "the memory and the wish occur in the same complex experience" would not entail "the memory and wish are simultaneous." The phrase "same complex experience" must mean an experience all of whose parts are simultaneous. But "simultaneous" must be taken here in a primitive absolute sense. It cannot, on pain of vicious circularity, mean "occur in the same complex experience of P." If it did, the same problem about whether all the parts of this experience were simultaneous or successive would arise again. Thus, "simultaneity" in the psychological sense reduces to nontemporal properties (e.g., belonging to a complex experience) and a primitive temporal property of metaphysical time, absolute simultaneity. This shows that every event in so-called psychological time occurs in metaphysical time.

To further buttress this point, consider the relation of "earlier" that belongs to so-called "psychological time." Suppose a headache is earlier than a backache in the sense that the headache is remembered by P and the backache perceived by P. But when do the remembering and perceiving occur? If the remembering occurs after the perceiving, that would not show the headache to be earlier than the backache. The remembering must occur *simultaneously* with the perceiving (or earlier than it), where "simultaneity" is meant in the sense of metaphysical simultaneity. Thus, metaphysical simultaneity must also belong to the reduction basis of the psychological relation of "earlier."

So far, I have argued that propositions and mental events exist in an absolute and primitive metaphysical time. But this does not yet touch the STR, since the STR purports to be a theory not about propositions or mental events but about physical events. Physical events, it might be said, are relatively temporally related even if mental events and propositions are not.

7.6 All Physical Events Are in Metaphysical Time

I take it to be an obvious fact that mental events are temporally related to physical events. For example, my wondering why I am alive is later than the physical event of my birth and earlier than the physical event of my death and bodily disintegration. And my sensation of pressure and warmth in my hand is (roughly) simultaneous with the physical event of my shaking somebody's hand.

Given this, there is an argument based on the premise that mental events are in absolute metaphysical time that shows that physical events are also terms of absolute relations. My act of thinking of the proposition *that a triangle is three-sided* is simultaneous with the events in my brain that correlate with this act of thinking (or if not, then with some other event in my brain). This relation cannot be physically relativistic, since there is no spatial midpoint between my thinking and the brain states at which physical emissions from both states could meet. And this relation cannot be psychologically relativistic, since there is no second mental act that coapprehends both my act of thinking the proposition and the brain events that correlate with it. Thus, these brain events must be absolutely simultaneous with these mental events and therefore with all the propositions and all the other mental events that are then absolutely simultaneous with these mental events.

The defender of the STR might respond that nothing I have said so far is inconsistent with the main relativist thesis of the STR, *that distant physical events sustain only relative temporal relations to each other*. The relativist thesis of the STR pertains to *distant simultaneity* and implies nothing about the relations among propositions, mental states, and brain states.

I respond that my arguments have implicitly established that distant physical events are absolutely related. An intelligent organism near the star Arcturus may be thinking of the proposition *that a triangle is three-sided,* and I may be thinking of it, as well, such that both of these acts of thinking are absolutely simultaneous with the proposition and thereby with each other. It follows that the brain states correlating to, and absolutely simultaneous with, these two acts of thinking are absolutely simultaneous with each other.

This might be denied. Call the event in my brain that is simultaneous with my thinking of the proposition about the triangle the event B1, and the other organism's brain event B2. And call my act of thinking the mental event M1 and the other's thinking the mental event M2. Now it might be argued by a defender of the STR that if M1 is absolutely simultaneous with B1, M1 with M2, and M2 with B2, that does not entail that B1 is absolutely simultaneous with B2. The two distant physical events B1 and B2 are only simultaneous relative to some reference frame.

But this argument is implicitly self-contradictory, since it is inconsistent with the analytic truth that absolute simultaneity is a transitive relation. It is analytically true that if A is simultaneous with B, and B with C, then A is simultaneous with C. Thus, if B1 is absolutely simultaneous with M1, and M1 is absolutely simultaneous with M2, then B1 is absolutely simultaneous with M2. Since M2 is absolutely simultaneous with B2, it follows that B1 is also absolutely simultaneous with B2.

But is it an analytic truth that absolute simultaneity is transitive? It seems to me that this is as self-evident as anything can be. The only semblance of coherence that may accrue to a denial of its transitivity derives from tacitly supposing that simultaneity is relative. I can suppose that A is simultaneous with B and B with C but that A is not simultaneous with C only if I suppose that these relations obtain relative to different reference points, for example, if I suppose that A is simultaneous with B, and B with C, relatively to a reference point R1 but that A is not simultaneous with C relatively to a reference point R2. Observe that this denial of the transitivity of simultaneity requires that I not only assume that simultaneity is relative but also change my reference points between (i) considering the first term's relation to the second term B and the second term's relation to the third term C and (ii) considering the first term's relation to the third term. If I retain the same reference point, R1, then "A is simultaneous with B, and B with C" entails "A is simultaneous with C."

However, a more general argument that shows distant physical events to be absolutely related is also possible. In fact, this argument shows that everything that actually exists is absolutely temporally related in metaphysical time. This argument is based on the temporal existence of propositions. Consider the *de dicto* proposition

(1) that the physical events E1 and E2 are distantly simultaneous relatively to the reference frame R1.

Assume that (1) is true, that is, that a relation of correspondence obtains between (1) and the state of affairs S that consists of E1 and E2's being distantly simultaneous relatively to R1. If we further assume two plausible principles, then it can be proven that the events E1 and E2 occur in absolute metaphysical time. The two principles are

(P1) If something A exists in the metaphysical time series T and possesses an n-adic property F, then A possesses F at some time in T.

(P2) If A's possession of F belongs to the time series T (i.e., occurs at some time in T), then A exists in T.

The proposition (1) exists in absolute metaphysical time; and, as I have assumed, (1) possesses the dyadic property of corresponding to the physical state of affairs S. By P1, (1) possesses this property at some time in absolute metaphysical time. At the time that it possesses this property, it is true of (1) *that (1) now corresponds to S,* where "now" refers to the time that is then present in the absolute time series. Another way to say this is *that there now [in absolute metaphysical time] obtains a relation of correspondence between (1) and S,* such that the dyadic property of correspondence is now coexemplified by (1) and S. But if this dyadic property *now* inheres in S, then S (by P2) also exists in absolute metaphysical time. To say that S exists in metaphysical time is to say that E1's and E2's exemplification of the distant simultaneity relation exists in metaphysical time; by (P2), it follows that the events E1 and E2 also exist in metaphysical time. If two events exist in metaphysical time, then they are related to each other by some of the temporal relations constitutive of metaphysical time. This implies that E1 and E2 bear some absolute temporal relation to each other; E1 is either absolutely earlier than E2, absolutely simultaneous with E2, or absolute later than E2. By a similar argument it can be shown that every event that belongs to STR time also belongs to metaphysical time and is absolutely temporally related to every other event that belongs to STR time. Indeed, similar arguments would also show that any event whatsoever belongs to metaphysical time and is absolutely temporally related to every other event and that all events occur in the same metaphysical time.

7.7 Metaphysical Time Is the Only Possible Time

I have argued that metaphysical time is primitive, absolute, and all-embracing. I have also argued that STR time and psychological time are reducible to nontemporal n-adic properties and the primitive n-adic properties of metaphysical time. If these arguments are sound, then metaphysical time meets the three conditions I set forth for being the only time. If we state conditions 2 and 3 in terms of STR time and psychological time, the three conditions may be stated as follows:

1. The events or states that are temporally related to each other in metaphysical time are related by primitive temporal relations, such that "earlier," "later," and "simultaneous" must be taken as indefinables in referring to these relations. The same holds for the A-properties.
2. The events that are related to each other in STR time or psychological time are related by reducible temporal relations, such that "earlier," "later," and "simultaneous" as referring to these relations can be defined in terms that contain no temporal words (except perhaps for words referring to the primitive temporal relations or metaphysical time). The same holds for the A-properties.
3. All the events that are related by the STR relations or the psychological time relations are also related by the primitive temporal relations of metaphysical time. The same holds for the A-properties.

My arguments imply that metaphysical time is the only time in the actual world and that it is the only time in any possible world in which there is time. This modal implication may be made explicit as follows. Consider that *de dicto* propositions exist in every possible world and thus in every world in which there is time. Since propositions exist in absolute metaphysical time in every world in which they exist in time, it follows that there is absolute metaphysical time in every world in which there is time. It is false that there are some possible worlds in which propositions exist in time but not in absolute metaphysical time, since if this were true, propositions would have to exist in some physical or mental time series in these worlds. They cannot exist in a physical time series, since they are spatially unlocated and cannot sustain physical relations to physical events. Moreover, they cannot exist merely in some mental time series. For example, it is false that there is some world in which propositions exist only in the mental time series of God or some other mind(s), since mental or psychological time is reductively definable in terms of absolute metaphysical time and psychological properties, as I argued in section 7.6. If something exists in a psychological time series, that is just a way of saying that it exists in absolute metaphysical time and stands in certain mental relations to certain mental events. Indeed, there is really no such thing as a psychological or physical *time* series, strictly speaking, since such alleged series are really series of mentally or physically related events that occur in primitive metaphysical time. Thus, absolute metaphysical time not only exists in every world in which there is time, but is the only time series in every world in which there is time.

Furthermore, there is *only one* absolute metaphysical time series in each world in which there is metaphysical time. This is proven by way of a reductio. Suppose there is an absolute metaphysical time series T and an existent E that does not exist in that time series. There is some other existent D that exists in T and is different from E. In other words, the relation of *difference* obtains between D and E. By the principle P1 stated in the last section, namely,

> (P1) If something A exists in the metaphysical time series T and possesses an n-adic property F, then A possesses F at some time in T,

this relation inheres in D at some time in T. Suppose that difference *now* (in T) obtains between D and E, such that this relation *now* inheres in both D and E. This entails, by P2 ("If E's possession of F belongs to the time series T, then E exists in T"), that E exists in the time series T; for if difference now inheres in E (where "now" refers to the present time in T), then E possesses a property at a time in T and thereby exists in T.

Note further that if E exists in T, then E does not exist in any other time series. If it did, it would exist at some times that are not simultaneous with, earlier than, or later than, any time in T. If it did exist at such a time—call it Tx—then its state at Tx would be different (at least numerically) from its states in T, such that a relation of difference would obtain between its state at Tx and its states in T. If this relation obtains now in T, then it follows, by an argument similar to the given one, that its state at Tx in fact occupies a time in T and therefore that the attempt coherently to describe a state of E that does not exist in T fails. (A similar argument would show that the time Tx is different from some time Ty in T and that Tx is therefore temporally related to Ty and by virtue of this temporal relatedness is really a time in T.)

These arguments entail that it is necessarily false that there are two or more parallel time series, that it is necessarily false that there is a branching time series, and that it is necessarily false that there is a branching–reconverging time series. However, it has been assumed as obvious by most twentieth-century philosophers of time that such multiple time series are possible. Accordingly, it might be thought that my argument just given is an altogether too rapid dismissal of the logical possibility of multiple time series. I suggest, however, that the expressions "parallel time series," and so on, as they are used in the theories of multiple time series, use "times" not in the sense I do but in the sense pertaining to Einstein's Theory of Relativity. It is, indeed, possible that there are many "time series" in the Einsteinian sense even if there is at most one "time series" in my sense. But strictly speaking, we cannot call Einsteinian time "time" but, rather, a nontemporal system, namely, a maximal system of light-connectable physical events. Manifestly, "there can be many maximal systems of light-connectable physical events" is consistent with "Necessarily, there is only one time series."

The claim that STR time is not, in fact, time at all and that metaphysical time is the only time amounts to the claim that the STR is false and the Nontemporal STR true (assuming the existence of the observations that have been taken to confirm the STR). The Nontemporal STR is the STR minus the theses that the physical relations it describes are temporal relations. A proponent of the Nontemporal STR will agree with a proponent of the STR that facts such as the following obtain:

(A) Two light signals arrive simultaneously at a place P1 from two distant events x and y, such that P1 is the midpoint between x and y relatively to the reference frame R1. These same two signals arrive successively at a different place P2 that is the midpoint between x and y relatively to the different reference frame R2.

The STR differs from the Nontemporal STR in that the former identifies these observable luminal relations with temporal relations, whereas the latter does not. The proponent of the Nontemporal STR will say of the events x and y that they are either absolutely simultaneous or absolutely successive but that he does not know which. The knowledge he possesses about x and y concerns not their temporal relationship but their luminal relationships; he knows that light signals sent from them intersect simultaneously at one relative midpoint between them but intersect successively at a different relative midpoint between them. The proponent of the Nontemporal STR will claim knowledge of the absolute temporal relations among his mental events (which he can directly observe), among propositions, and between propositions and his mental events. He will also claim (approximate) knowledge of the temporal relations among local physical events, between local physical events and his mental events, between local physical events and propositions, and between any of these and the mental events occurring in minds that are locally embodied. But he will not claim knowledge—even approximate knowledge—of the temporal relations among distant physical events that are connected by the STR relations of "distant simultaneity" or "distant succession."

But this is not to say that the proponent of the Nontemporal STR has no knowledge of the temporal relations among distant physical events. He does have temporal knowledge of the distant physical events that are related by the STR relations of *being topologically earlier than* (being in the absolute past of) and *being topologically later than* (being in the absolute future of). Event E1 is topologically earlier than E2 if and only if there can be a luminal or subluminal causal influence emanating from E1 that affects E2, relative to all reference frames. If E1 is causally related to E2 relative to all reference frames, then given that a cause precedes its effect, E1 is *earlier* (in metaphysical time) than E2. The proponent of the Nontemporal STR lacks knowledge merely of the temporal relations among events that are connected by the STR relation of topological simultaneity, for only these events sustain the relativist STR relations of distant simultaneity or distant succession.

7.8 Metaphysical Time Distinguished from Other Candidates for "Absolute Time"

The conclusions about metaphysical time and the STR reached in the preceding sections should not be confused with superficially similar theories that have been advanced. In order to avoid some of these confusions, I shall distinguish my concept of metaphysical time from (i) cosmic time, (ii) physically reducible absolute time, (iii) neo-Lorentz time, (iv) Newtonian time, and (v) divine time.

A. Cosmic Time

The phrase "cosmic time" is standardly used to denote the time of the universe as a whole as it is represented in the General Theory of Relativity (GTR). Cosmic time is based on the notion of a spacelike hypersurface that is orthogonal to the time axis of a reference frame; for example, in Figure 7.3, the x-axis is a spacelike hypersurface orthogonal to the time axis t at the event A. Consider now a set of interlocking hypersurfaces that are orthogonal to the parallel time axes t, t', t", . . . at the events A, B, C, . . . such that A is distantly simultaneous with B, B with C, and so on. The plane of simultaneity formed by these interlocking hypersurfaces can be extended to form a universewide plane of simultaneity. We may then form a sequence of all the universewide planes of simultaneity that are orthogonal to these time axes, ordered by the relation of *earlier than* and identify this sequence with cosmic time.[10]

Now, there are many different ways to cross-section the universe into serially ordered planes of simultaneity; there is one cross-sectioning for each set of time axes that are nonparallel to all other sets of time axes. Each different way of cross-sectioning the universe is a different cosmic time series. But if the universe is homogeneous and isotropic (matter is evenly distributed, and from each point the universe looks the same in every direction) then the simplest way to cross-section the universe is in terms of the *planes of homogeneity*. Each plane of homogeneity is

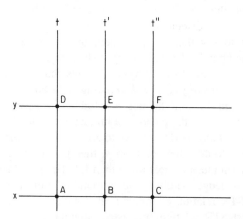

Figure 7.3 Cosmic Time. A, B, and C are events that are located on the time axes t, t', and t", respectively. The spatial axis x is a set of interlocking spatial hypersurfaces that are orthogonal to the time axes t, t', and t" at the events A, B, and C. The x-axis can be maximally extended to the left of A and to the right of C to form a universe-wide plane of simultaneity. This plane of simultaneity is earlier than the plane of simultaneity y, such that x and y are two successive planes of simultaneity that belong to cosmic time.

a universewide set of interlocking hypersurfaces such that at each position on one of these hypersurfaces, the density and pressure of matter and the curvature of the universe are the same. Since our universe is approximately homogeneous and isotropic, cosmologists use the cosmic time series that consists of the successive planes of homogeneity as the temporal framework from which to date events. (I would note parenthetically that the A-expressions in the cosmological writings I quoted in section 1.5 all refer to A-positions in the cosmic time series consisting of the planes of homogeneity.) Let us call this particular cosmic time series simply "cosmic time," since this is the only cosmic time series that is standardly represented in current cosmological theory. The question that we shall ask is whether the planes of simultaneity of cosmic time are *absolute* simultaneity relations and whether they can be identified with those of metaphysical time.

Richard Swinburne believes that the planes of simultaneity of cosmic time are absolute simultaneity relations. After presenting the theory of cosmic time, he writes: "So I conclude that Newton is right about time. There is in his sense Absolute Time."[11] But this is a mistake, since the simultaneity relations of cosmic time are *relative* to the reference frames whose time axes are orthogonal to the planes of homogeneity. The statement that "x is simultaneous with y," where "simultaneity" is used in the sense of cosmic time, is elliptical for "x is distantly simultaneous with y *relatively to a reference frame whose time axis is orthogonal to a plane of homogeneity.*"

This entails that the simultaneity relation of cosmic time is not the simultaneity relation of metaphysical time. A second difference is that cosmic simultaneity is not, in fact, a temporal relation at all, since it is reducible to nontemporal relations and properties and the primitive relation of simultaneity that belongs to metaphysical time. Cosmic simultaneity reduces to the n-adic properties that compose distant simultaneity, as well as the property of *being a plane of homogeneity*. (But the distant simultaneity of the STR is not quite the same as the distant simultaneity pertinent to the GTR, since the STR notion involves inertial reference frames in Minkowski space–time, whereas the GTR notion involves, instead, reference frames that are inertial trajectories of freely falling particles in Riemannian space–time.) This shows that cosmic time is not really *time* but, instead, is a sequence of causally connected planes of homogeneity.

B. Physically Reducible Absolute Time

The concept of physically reducible absolute time is often the notion used by philosophers of science when they wish to oppose the STR to an alternate theory of time. According to the physically reductive theory of absolute time, there is a relation of absolute (dyadic) simultaneity but that this is not a primitive temporal relation but is definable in entirely nontemporal terms. Absolute simultaneity is defined as the physical relation of *connectibility by an infinitely fast signal*, where the *defiens* contains no temporal words at all. I would say that if such a relation obtained between two physical events, the events would be absolutely simultaneous in the primitive sense of metaphysical time but that if two events are absolutely simultaneous in the primitive sense, they need not be connectable by an infinitely fast signal. One reason for this is that some absolutely simultaneous events are nonphysical. Another is that physical events can be absolutely simultaneous even if there are no infinitely fast signals. Perhaps infinitely fast signals are needed if two distant events are to be *observed* (or "verified") to be absolutely simultaneous (in the primitive sense), but the existence of these signals is not logically required by the obtaining of this simultaneity relation between distant events.

A different concept of physically reducible absolute time is based on a certain interpretation of the outcome of Alain Aspect's[12] verification of Bell's theorem. One may interpret Aspect's experiment as showing that a causal influence is instantaneously transmitted between two distant photons, such that their polarizations are made to correlate. Since these causal transmissions are not *signals,* a reductive definition of absolute simultaneity in terms of *connectibility by these infinitely fast causal transmissions* would be different from the aforementioned definition. Nevertheless, it would be a physically reductive definition of absolute simultaneity. Since the obtaining of this physical relation entails the obtaining of the primitive simultaneity relation of metaphysical time, we may take Aspect's experiments as indicating that there are physical manifestations or evidence of the absolute temporal relations of metaphysical time.

C. Neo-Lorentz Time

Neo-Lorentz time is postulated by such contemporary physicists as H. E. Ives, Geoffery Builder, S. J. Prokhovnik, and Martin Ruderfer.[13] The most serious problems with Lorentz's original theory were eliminated by H. E. Ives, who produced a theory that avoided the ad hoc nature of Lorentz's and derived the Lorentz transformation equations from (a) the laws of conservation of energy and momentum and (b) the laws of transmission of radiant energy. According to the neo-Lorentz theory, there is a privileged reference frame R that is at rest in absolute space, and true time is the time measured in R. We are not able to experimentally identify which of the many reference frames is R, so we have no access to true time, only to the apparent times of the different reference frames.[14] But it is nonetheless the case that there is a true time. The relations of distant simultaneity and distant succession that are relative to R are the temporal relations that really obtain among events, and the relations that are measured by other reference frames are merely apparent temporal relations.

Neo-Lorentz time consists of the relations of distant simultaneity and succession that are relative to a reference frame at rest in absolute space, where "distant simultaneity" and "distant succession" are defined in terms of light signals intersecting at the absolute spatial midpoint between the events. This implies that neo-Lorentz time is no less physically reductive than STR time. Furthermore, the neo-Lorentz temporal relations would be relative, even though privileged (being the only real temporal relations); for they would be relations that events possess relatively to a reference frame at rest in absolute space. In these two respects, they would differ from the relations constitutive of metaphysical time. But the physical relations that would obtain in this privileged reference frame would count as physical evidence for the primitive temporal relations of metaphysical time. The statement that "E1 and

E2 are distantly simultaneously relatively to a reference frame at rest in absolute space" would be evidence for "E1 and E2 are absolutely simultaneous in metaphysical time."

D. *Newtonian Time*

Metaphysical time also differs from Newton's absolute time. But the reason for this is not that Newton's absolute time is a physically reducible time. Unlike the three conceptions just considered, Newton's time is not definable in terms of physical relations among events. This is the point of Newton's distinction between absolute time ("duration") and relative time in the scholium to his Definitions in the *Principia*. He states that absolute time "from its own nature flows equably without regard to any thing external" and that "Relative, apparent, and Common Time . . . is some sensible and external (whether accurate or unequable) measure of Duration by the means of motion, which is commonly used instead of True time." Nor does the difference between Newtonian time and metaphysical time consist in the fact that simultaneity in Newtonian time does not entail simultaneity in metaphysical time; there is such an entailment.

The differences are found elsewhere. One difference is that metaphysical time is a time in which propositions, mathematical states of affairs, and other abstract objects exist; but this is not the case for Newtonian time. Furthermore, the justification for believing there is metaphysical time is that some concrete or abstract object acquires or loses some n-adic property ("Something exists in metaphysical time if and only if it acquires or loses any n-adic property"); but this is not the justification that Newton offered (or would offer) as sufficient for the postulation of his absolute time. Rather, there are three physical justifications for Newtonian time, pertaining to his laws of motion and gravitation. These justifications are

1. *The obtaining of Newton's law of gravitational attraction.* According to this law, gravity is a force that propagates instantaneously across arbitrarily large distances. This propagation requires the emission of the gravitational force to be absolutely simultaneous with the effect of the emitted force upon the distant body.
2. *The obtaining of Newton's first law of motion.* According to this law, a body on which no force is acting either remains at rest or moves uniformly in a straight line. The uniformity of this motion requires equal intervals of uniformly passing time, such that these intervals are related by absolute relations of earlier and later.
3. *The obtaining of Newton's second law of motion.* This law implies that particles can be accelerated to arbitrarily high speeds by impressed forces. This implies that if a particle can be accelerated to a speed arbitrarily close to an infinitely fast speed, then the particle can come arbitrarily close to being at two places at the same time. This requires the notion of the absolute simultaneity of the two places.

Newton would not have regarded the postulation of his absolute time as physically justified if these three conditions had not obtained. However, the postulation of metaphysical time would be justified (i.e., by the arguments advanced in the preceding sections) even if these three conditions did not obtain. It is clear that metaphysical time can obtain even if these three conditions do not obtain. There is no logical contradiction in supposing there is absolute metaphysical time and yet that gravity is not an instantaneous force, that it is not the case that unimpressed bodies remain at rest or in uniform motion, and that particles cannot be accelerated to arbitrarily high velocities.

But the most important difference between metaphysical time and Newtonian time is that the latter is partly reducible to nontemporal n-adic properties and hence is not really time at all. Newtonian time is divine time, which reduces to metaphysical time and mental relations in God's mind. The general scholium to the *Principia* added in 1713 states that absolute time is an attribute of God: God "endures forever, and is everywhere present; and, *by existing always* and everywhere, *he constitutes duration* and space" (my italics). Newton does not spell out exactly what he means by saying that God *constitutes duration;* but his remarks suggest a reductionist theory of divine time, which I shall examine in the context of William Lane Craig's more developed version of this theory.

E. Divine Time

William Lane Craig's theory of divine time is in part based on a critique of the verificationist philosophical basis of the STR. Craig writes:

> I find it surprising that anyone reading Einstein's 1905 paper can think that Einstein *demonstrated* that absolute simultaneity does not exist and that time is therefore relative to a reference frame. For the entire theory depends upon acceptance of Einstein's arbitrary (and, indeed, highly counter-intuitive) definition of simultaneity, coupled with a philosophical positivism of Machian provenance according to which a notion like absolute simultaneity is meaningless if it is empirically undetectable. . . . One who is not a positivist and who therefore rejects Einstein's definitions would regard these relatively moving observers as deceived due to the nature of their measurements, which fail to detect true time. In a real sense, he would not regard Einstein's theory as a theory about time and space at all, but, as Frank put it, as "a system of hypotheses about the behavior of light rays, rigid bodies, and mechanisms, from which new results about this behavior can be derived." Trapped in our locally moving frames, we may be forced to measure time by devices which are inadequate to detect the true time, but that in no way implies that no such time exists.[15]

If one rejects verificationism, one is deprived of the main reason Einstein implicitly offered to accept his thesis that time is the observable system of relativistic luminal relations. However, the rejection of verificationism does not entail that there is absolute time; it merely entails that Einstein's reason for rejecting it is untenable. There may be other reasons to accept the STR theory of time; one could argue on grounds of *simplicity* or *parsimony* that the STR is preferable to a theory of absolute time. Thus, if one is to believe in absolute time, some positive argument is needed, not simply a rejection of verificationism. Although Craig does not offer a positive argument in the quoted passage, in other passages he claims that the belief in an absolute time is justified by the belief that God exists and is in time. He writes, "God's time is the true A-series time, determined by the succession of events in the divine consciousness and activity."[16]

But this divine time is not metaphysical time, since it is neither primitive nor absolute. To begin with, time in the sense of a series that consists of, or is determined by, a succession of events in God's mind is a *psychological or mental time* and as such is not *time,* that is, the primitive time that is not reductively definable in terms of psychological events and n-adic properties. The psychological relations among the mental events in God's mind are no more *temporal relations* (in the primitive sense) than are the physical or luminal relations among physical events. It is rather the case that the psychological relations among God's mental events and the physical relations among physical events obtain *in* time, that is, in time in the nonreductionist sense.

There is a further problem with Craig's theory, pertaining to his claim that time is absolute. I do not believe that his divine time is an absolute time, because "E1 and E2 are simultaneous," by his theory, means that E1 and E2 are simultaneous relatively to God. Craig writes that "*From God's perspective* in real, A-series time, there is an absolute present. . . . There must be an absolute, cosmic 'now' which describes the state of the universe *as it is present to God.*"[17] Presentness in divine time is the relational property of *being present from God's perspective* and simultaneity in divine time is *being simultaneous from God's perspective.* The simultaneity relation would not be dyadic (absolute) but triadic (relative) and would have God for its third relational term.

Divine time is not only reducible and relative but is also a dubious postulate. Craig's case for his "true and absolute time" is precarious because it depends on the claim that God exists. There is divine time only if God exists, and God exists only if there is a solution to the problem of evil; however, there appears to be no solution to this problem in sight.[18]

Notes

1. For references, see nn. 2–4.

2. Robert Weingard, "Relativity and the Reality of Past and Future Events," *British Journal for the Philosophy of Science* 23(1972): 119–21; Ferrell Christensen, "Special Relativity and Space-like Time," *British Journal for the Philosophy of Science* 32(1981): 37–53, esp. p. 46; D. H. Mellor, "Special Relativity and Present Truth," *Analysis* 34(1974): 74–78; Nicholas Maxwell, "Are Probabilism and Special Relativity Incompatible?," *Philosophy of Science* 52(1985): 27; and Paul Fitzgerald "The Truth About Tomorrow's Sea Fight," *Journal of Philosophy* 66(1969): 307–29.

3. Storrs McCall, "Objective Time Flow," *Philosophy of Science* 43(1976): 337–62; Lawrence Sklar, *Space, Time, and SpaceTime* (Berkeley: University of California Press, 1974), 272–75; Fitzgerald, "The Truth About Tomorrow's Sea Flight."

4. Howard Stein, "On Einstein–Minkowski Space–Time," *Journal of Philosophy* 65(1968): 5–23; E. Harris, "Simultaneity and the Future," *British Journal for the Philosophy of Science;* D. Dieks, "Special Relativity and the Flow of Time," *Philosophy of Science*

55(1988): 456–60; Maxwell, "Are Probabilism and Special Relativity Incompatible?"; Mellor, "Special Relativity and Present Truth."

5. Putnam's interpretation is distinct from these three interpretations, but his interpretation is meant to establish not the consistency but the inconsistency of the tensed theory of time and the STR. See the chapter "Time and Physical Geometry" in his *Philosophical Papers,* vol. 1 (Cambridge: Cambridge University Press, 1975).

6. Herman Bondi, *Relativity and Common Sense* (New York: Dover, 1964), 168.

7. For a summary of this evidence, see Craig's forthcoming book on God and time.

8. Hans Reichenbach, *The Philosophy of Space and Time* (New York: Dover, 1957), p. 124.

9. Some of these difficulties are explored in Quentin Smith, "Atheism, Theism, and Big Bang Cosmology," *Australasian Journal of Philosophy* 69(1991): 48–66; idem, "A Big Bang Cosmological Argument for God's Nonexistence," *Faith and Philosophy*, 9(1992): 217–237; idem, "An Atheological Argument from Evil Natural Laws," *International Journal for the Philosophy of Religion* 29(1991): 159–174.

10. For further discussion of cosmic time, see my "Natural Explanation of the Existence and Laws of Our Universe," *Australasian Journal of Philosophy* 68(1990): 22–43.

11. Richard Swinburne, *Space and Time*, 2d ed. (St. Martin's, 1981), 202.

12. See Alain Aspect and Phillipe Grangier, "Experiments on Einstein–Podolsky–Rosen–type Correlations with Pairs of Visible Photons," in *Quantum Concepts in Space and Time* (Oxford: Clarendon, 1986), 1–15.

13. Herbert Ives, "Derivation of the Lorentz Transformations," *Philosophical Magazine* 36(1945): 392–401; Martin Ruderfer, "Introduction to Ives' 'Derivation of the Lorentz Transformation'," *Speculations in Science and Technology* 2(1979): 243; S. J. Prokhovnik, *Light in Einstein's Universe* (Dordrecht: Reidel, 1985); G. Builder, "Ether and Relativity," *Australian Journal of Physics* 11(1958): 279–97. For a philosophical discussion of the neo-Lorentz theory, see William Lane Craig's forthcoming book on God and time.

14. William Lane Craig argues, however, that we are able to verify which reference frame is R; R is a frame whose time axis is orthogonal to a plane of homogeneity. This is argued in his forthcoming book on God and time.

15. William Lane Craig, "God and Real Time," *Religious Studies* 26(1990): 335–47.

16. Ibid.

17. Ibid.

18. See n. 9.

8

Conclusion

8.1 Summary of the Argument for Presentism

In Chapter 7 I argued that metaphysical time is primitive, absolute, all-embracing and is the only time series, all other time concepts being reductively definable in terms of nontemporal n-adic properties or in terms of these properties and the primitive temporal properties of metaphysical time. If these arguments are sound, then they justify one of the theses of presentism, namely, that presentness inheres in events absolutely, not relatively to a reference frame (or mind).

The theory of presentism for which I have been arguing in this book may be summarized in one sentence as the thesis that "all states of affairs are of the form *presentness-inheres-in-such-and-such*," where "inheres-in-such-and-such" is understood in a broad sense so that this form includes such states of affairs as

1. Presentness inheres in John's walking.
2. Presentness inheres in the pastness of Plato's thinking.
3. Presentness inheres and will always inhere in the simultaneity of an event E on Earth and an event E' in the Andromeda galaxy.
4. Presentness inheres, and always has inhered, and always will inhere in the equality of two to one plus one.

The summary presentist thesis that "all states of affairs are of the form *presentness-inheres-in-such-and-such*" entails the specific theses for which I have argued in both parts of this book:

This summary thesis entails that the tensed theory of time is true, for states of affairs are of this form only if A-sentences ascribe A-properties to events and only if some tokens of these sentences are true. Furthermore, the presentist thesis entails that the "no-property tensed theory of time" of Prior and others is false, since if there is no property of presentness, then no state of affairs is of the form presentness-inheres-in-such-and-such. The presentist thesis entails, additionally, that McTaggart's paradox does not provide a sound argument that there is a vicious infinite regress involving the inherence of presentness in events or moments.

The presentist thesis also entails that all so-called tenseless sentences, such as

251

B-sentences, natural law sentences, and analytic sentences are in fact multiply tensed, that is, are synonymous with sentences that contain more than one tensed copulae, such that the so-called tenseless "is" is synonymous with expressions such as "is and will always be" or "always was, is, and always will be." This is tantamount to the claim that all so-called "tenseless sentences" have presentness for a logical subject; for by virtue of having a multiply tensed copulae, these sentences refer to presentness and ascribe it a property of a form such as *inheres and always will inhere in such-and-such*. This entails, in turn, that all the states of affairs that correspond to true tokens of these "tenseless sentences" have the form presentness-inheres-in-such-and-such (understood in the aforementioned broad sense).

The thesis that all states of affairs have the form presentness-inheres-in-such-and-such (and do not have the form presentness-inheres-in-such-and-such-*relatively-to-a-reference-frame*) also entails that presentness inheres in events absolutely and therefore that the STR is not a true theory of the nature of time.

If presentism is true, then there is a uniform structure to reality in the sense that each state of affairs consists of presentness as possessing a property of the form *inhering-in-such-and-such*. In this sense, all of reality is unified in presentness. This suggests a sense in which presentness is the most distinctive and preeminent property, namely, that it is the only property that is a metaphysical subject of every state of affairs. Indeed, in this sense, it is the preeminent object or existent in general; for it is the only concrete or abstract object that is the metaphysical subject of every state of affairs. By virtue of being the "preeminent object" in some sense, the property of presentness occupies in presentism a status analogous to that occupied by the property of unity ("the One") in Neoplatonism, although of course there are obvious differences (e.g., all of reality does not "emanate" from presentness). Accordingly, we may say with truth—albeit metaphorically—that presentism is the glorification of presentness, in the sense that it entails that all of reality is a circle whose center is presentness and whose radii are properties of the form inheres-in-such-and-such.

8.2 Ethical Implications of the Argument for Presentism: The Presentist Attitude

It is one thing to understand the argument that reality is unified in presentness, but it is another to affectively experience this unification. The affective experience may be called *the presentist attitude;* this is an experience wherein the person's psychological unity mirrors the unity of reality. The affective experience of the unity of reality is one of the human excellences, along with aesthetic appreciation, the attainment of scientific knowledge, the attainment of philosophical knowledge and the like. According to perfectionist ethics, the promotion of these excellences is the most important ethical goal of human life and according to other ethics, their promotion is one of the ethical goals of human life. The human excellence of understanding the argument for presentism is an excellence pertaining to the attainment of philosophical knowledge, but there is a different human excellence that pertains to the affective experience of the unity of reality and this latter excellence pertains to the presentist attitude. The affective experience of the unity of reality is of distinct value in its own right and compares in kind with Plato's vision of goodness itself, Plotinus's vision of the property of oneness, Spinoza's intellectual love of Nature, Heidegger's dread of the Being of beings, and Munitz's awe of Existence. The discovery of these valuable experiences is not the province of ethics but of metaphysics, since the discovery requires a metaphysical investigation that determines which experience is the true experience of the fundamental structure of reality. The implication of the argument for presentism is that the value accruing to the experience of the basic structure of reality is a value possessed by the presentist attitude rather than by some other attitude. The presentist attitude is the supervenience base upon which this value supervenes. The nature of the present attitude will be analyzed in detail in the following pages.

The presentist attitude involves a type of speaker's reference or, more generally (to cover cases of listening, writing or mentally tokening a sentence), *language-user's reference*. Language-user's reference is a triadic relation of the form "P refers to x *via* t," where P is a person, x the concrete or abstract object referred to, and t some token of a phrase, word or morpheme. The token t stands in the relation of semantic reference to the object x, but the relation in which the person stands to x is a psychological or mental relation, the relation of *intentional reference;* the person is conscious of x, whereas the token is not. But since the intentional references we are discussing are language-using references, the intentional references are to the object x *as* the semantic referent of the token t. Thus, if Alice refers to presentness when she utters "John is running," she intentionally refers *to* presentness *as* the semantic referent of her utterance-token of the grammatically present tensed form of the copula "is." The intentional reference to presentness is a *universal language-user's reference* in that every linguistic act includes a reference to presentness. A linguistic act involves tokening a sentence (orally, inscriptionally or mentally) and experiencing a propositional attitude to the proposition expressed by the sentence-token. Just as every sentence-token has presentness for a logical subject, so every linguistic act includes an intentional reference to presentness. Other language-user's references are merely *local reference-acts;* for example, the reference to John or running is local in that this intentional reference is a part of some but not all linguistic acts. Alice locally refers to John as the semantic referent of "John" and running as the semantic referent of "running."

In addition to language-user's references, there are language-user's predications. Alice does not simply mentally refer to a series of objects; she mentally *predicates* running of John and presentness of John's running. Mental predication or "language-user's predication" parallels semantic predication (the semantic relation of *conveying that* parts of the proposition are propositionally E or C related), just as mental reference parallels semantic reference.

Our normal linguistic acts are not presentist attitudes. A presentist attitude is an attitude to presentness as inhering in all of the infinite number of beings that there are. The one infinite presence comes to explicit appearance in this attitude. A presence is an inherence of presentness in some being and the infinite presence is the inherence of presentness in the infinity of beings that there are. In a presentist attitude I attentionally refer to presentness and predicate of it the property of inhering in the infinity of beings. Our normal linguistic acts are not presentist attitudes if only for the reason that they are not attitudes to presentness and are not attentional references to presentness. This is reflected in the normal syntactical form of sentences. Usually, a concrete proper name or definite description is the grammatical subject of our sentences, as in "John is attacking me." The grammatical subject has the pragmatic or psychological property of *directing attention upon its referent, in such a way that its referent becomes the accusative of our affective attitudes* (what we have an attitude to). We refer to presentness only as a part of our predicative awareness, our awareness of the semantic content expressed by the grammatical predicate ("is attacking me"), such that we do not have an attitude to presentness, but rather grasp presentness only as a part of the complex property, *presently attacking me,* that we predicate of the accusative of our attitude and take as the reason for our attitude to this accusative.

The first step in realizing the presentist attitude must accordingly be a *reversal* of one's normal attitude. Instead of experiencing an attitude to a first order substance or event, we shift our attention to presentness and experience an attitude to the *presentness* of the substances or events, or to the *presentness* of their pastness or futurity. This reversal involves relegating the first order substance or event to a part of that of which I have a predicative awareness. The person or tree or storm no longer appears as the accusative of my attitude but appears adjectivally, as a part of a property I predicate of presentness and take as a reason for my attitude to presentness. The first order existent to which I had been experiencing an attitude now appears as a part of a second order existent, a part of a property of the form *inheres in B,* where "B" stands for a being in the present tensed sense ("a being" in the present tensed sense refers to something in which presentness inheres). I now grasp the sky only as part of the property *inhering in the blueness of the sky,* which is a property of presentness and thus a second order property.

Attitude reversal can be further characterized in intuitively evocative terms if we indicate that it involves letting go of the physical and personal things or events upon which we are normally focused and letting our attention swing onto *no-thing,* the void of presentness. My attention comes to rest on the very persisting of things, their being present and remaining present. I grasp their pure "being-there-at-all-ness," that is, their being present rather than their being past or future or never being at all. But this way of putting it is somewhat misleading since I do not grasp the void of presentness as adjectival upon things but things as adjectival upon the void. The "being-there-at-all-ness" is the metaphysical subject that possesses *inhering in the valleys below* and *inhering in the blue sky above.*

Presentist sentences can be used to achieve this attitude reversal. Shifting my attention from one part of a proposition to another part is realized by means of taking the latter part as the referent of the grammatical subject of some sentence that expresses the proposition. I choose to express the proposition I am grasping by "Presentness inheres in the blueness of the sky" rather than by "The sky is blue." The presentist structure of the proposition is now made syntactically explicit, and this in turn serves to render the presentist structure of the proposition, and of the corresponding state of affairs, explicitly apparent to me.

Attitude reversal involves a shift in my emotional attitude. Affective or emotional attitudes are attitudes *to* something *because of* something, where the latter something is the reason for the attitude. For example, I admire Jane because of her artistic accomplishments. Emotional attitudes correlate to language-user's reference and predication in that the emotional attitude is *to* the attentional language-user's referent *because of* the property that the language-user attentionally predicates of this referent. If I gaze out the window of the plane in a normal emotive frame of mind, I may experience an enchantment with the moon because of its brightness. But if I reverse my attitude by mentally tokening "Presentness inheres in the brightness of the moon" my emotional "aboutness" will shift from the moon to presentness. I will be enchanted with presentness because it inheres in the brightness of the moon. I am now appreciating the universal metaphysical subject, presentness, rather than one of the many local metaphysical subjects. This involves not just a change in the "aboutness" of the emotion but also in its quality; it changes from an ordinary enchantment to a *sublime* enchantment. Sublimity is a simple and indefinable generic feeling-quality in the experience of which I *feel* elevated above of the ordinariness of day to day first order living (absorption in this or that or this other first order existent) to the loftier plane of contemplation of the universal subject. If I am now appreciating first order reality at all, it is from the point of view of its transcendent unity, the unity of first order existents that transcends and encompasses them (transcends them because it is a higher order existent and encompasses them because it inheres in them).

The sublime modification of my feeling-quality is accompanied by a detachment from the existent(s) I had been appreciating. A different example illustrates this most instructively. Suppose a family member or person close to me N dies and I feel grief. I am grieving for N because he or she has passed away. But I wish to experience attentionally the universal reference-act and to experience it as it is related to this instance. I elicit this experience by mentally tokening "Presentness inheres in the pastness of N." I am no longer feeling grief about N but about presentness because of its inherence in N's pastness. I am now appreciating presentness rather than N and this detaches the "aboutness" of my emotion from N and attaches it to presentness. My grief changes to become a sublime sorrow about the universal subject, a sorrow that it no longer inheres in N. This detachment can be realized for any ordinary emotion I am experiencing, so that it is possible for all my emotions to be experienced as about one and the same metaphysical subject. As long as I realize this detachment, nothing in the ordinary world will move me since I am attached only to the universal metaphysical subject. Psychologically, it may be said that my emotional life becomes unified around presentness rather than parceled out among the many local metaphysical subjects. My emotions are now one and all responses to presentness in respect of its inherence in beings, rather than responses to these variegated beings themselves in respect of their properties.

But attentionally and sublimely referring to presentness is not sufficient to realize a presentist attitude. A presentist attitude is one in which the infinite presence comes to explicit appearance. The infinite presence is presentness's inherence in the infinite number of beings that there are. If we use "being" in a present tensed sense to refer to something in which presentness inheres, we may say that every item is either a being or a part of a being. Since John is present, John is a being and since Dante is a part of something in which presentness inheres, the state *the pastness of Dante,* Dante is a part of a being. Accordingly, if I take presentness as the accusative of my affective attitude and take *inhering in every being* as the reason for my affective attitude to presentness, I am experiencing a presentist attitude.

Any linguistic act consists of both mental reference-acts and predicative-acts. All the reference-acts that are not references to presentness are references to the being, or to a part of the being, which belongs to the property *inheres in B* that helps to make up the corresponding state of affairs (if the relevant proposition is true). Consider now the conjunction of all states of affairs. Since presentness is the metaphysical subject of each of these states of affairs, the conjunction of all the states of affairs is the *super-state of affairs* consisting of presentness as inhering in B1 and B2 and B3, . . . and so on for every being in which presentness inheres. This property has absolutely infinite conjuncts, where "absolutely infinite" is used in Cantor's sense to mean that for any transfinite cardinal number N, the number of conjuncts is more N. This absolutely infinite property of presentness is the *super-property.* Since presentness inheres in some new being at each new instant, present-ness possesses a new super-property at each new instant.

At any time t, whatever can be referred to at that time is either presentness, the super-property presentness than possesses, or some part of this super-property. This experience of the infinite presence is a knowing of the unity of reality, that every state of affairs is unified in presentness in that every state of affairs is a state of one and the same existent, presentness. By knowing that presentness possesses a super-property, I am knowing that every state of affairs is a state of presentness, that is, that every state of affairs is of the form, presentness-as-inhering-in-B.

Since my propositional attitudes are temporally extended rather than instan-taneous, my presentist experience is expressed by a sentence that reflects this extension, namely,

(1) Presentness successively possesses different super-properties.

But experiencing a propositional attitude to the proposition expressed by (1) is not all there is to the presentist attitude. If I merely grasp the proposition expressed by (1), I single out presentness but conceive merely generally the absolute infinity of beings in which presentness inheres. A more explicit experience of the infinite presence and the unity of reality is achieved if I single out as many beings as I possibly can. This is achieved in the *maximally expanded mental state.* This re-quires I do not merely cognize all states of affairs but single out as many as it is possible for me to single out and conceive generally all the rest. The distinction between singling out and generally conceiving all states of affairs is exemplified by the distinction between comprehending *that presentness inheres in John and in Alice, who both are standing in the room* and comprehending *that presentness inheres in whomever is standing in the room;* in the former comprehension I single out John and Alice but in the latter comprehension I conceive them generally. The more states of affairs I single out rather than conceive generally the more expanded my awareness, since I am aware of more of the details of reality. There are many ways to single out states of affairs, but the way in which the largest number of states of affairs can be singled out in a single act of comprehension is through outer and inner perception, where I perceptually grasp the states of affairs constitutive of my surroundings and of my current psychological state. This perceptual singling out is maximalized if I grasp my complete *indexical location,* the location that corresponds to a use of "I am here now" or, in presentist language, "Presentness inheres in my being here and now." Tokening this presentist sentence is instrumental in my becoming aware of presentness as inhering in myself and my experiences (bodily feelings, affective sensations, mental images, acts of awareness, etc.). In connection with this awareness I also become aware in external perception of presentness as inhering in my physical surroundings, my spatial location—the *here* that encompasses everything within my perceptual horizon, perhaps just my room or perhaps an entire mountain range that unfolds before my eyes. *Now* refers to the moment of time that is present and I become aware of presentness as inhering in this moment. The complex property ( ) *inhering in my being here and now* that is possessed by presentness is a conjunct of the super-property, the conjunct that constitutes my complete indexical location. This conjunct as possessed by presentness is a state of affairs, such that this state of affairs is itself a conjunct of the super-state of affairs, the super-state of affairs being the conjunction of all states of affairs.

The maximal expansion of my awareness involves a general conception of all other states of affairs, but a conception that is as specific as possible. This is achieved by conceiving all other states of affairs in the respects in which they are similar to the indexical states of affairs I single out. Corresponding to my singling out of myself as something in which presentness inheres, I am aware of presentness as inhering in *everyone else* besides myself, where "everyone else" is used in a broad sense to refer to other humans and to all other animals and, indeed, to everything that has experiences of some sort, regardless of the level of dimness or primitiveness. If some current speculations by physicists are correct that there are an infinite number of other intelligent organisms in the universe on an infinite number of other planets, then "everyone else" includes them in its scope as well. If there are disembodied persons, then they too are included in the extension of "everyone else." I expand my awareness not only to include presentness's inherence in everyone else but also in *everywhere else* besides here, or all other spatial locations. This not only includes all other spatial locations spatially connected to this one but the spaces in all other spatially disconnected universes, if there are any. This involves a grasp of presentness as inhering in whatever occupies these spatial positions as well. Third, I become aware of presentness inhering in the pastness or futurity of *every other time* besides the present one and the pastness or futurity of whatever occupies these times.

Mentally tokening certain presentist sentences is instrumental in building up to this maximally expanded awareness. The first stage is to token "Presentness inheres in myself, here and now" and attentionally realize an attitude to presentness as inhering in my complete indexical location. From here I go on to token "presentness also inheres in everyone else, everywhere else and in the pastness or futurity of all other times and their occupants," which enlarges my awareness in such a way that I achieve a general attentional awareness of all beings that lie beyond my indexical location. Given the maximal expansion of my awareness induced by these sentence-tokenings, I may then token "presentness is successively possessing different super-properties" in a way that expresses the maximally explicit experience of the unity of reality. This latter sentence may be tokened in a way that does not express this maximally explicit experience and for this reason the prior tokenings are required to ensure the maximally explicit experience is the one expressed. The sentence in question may be used to express a purely general attitude to super-states of affairs, an attitude that does not single out any states of affairs, and this use of the sentence does not express the maximally explicit experience. But if I have a built up a maximally expanded awareness by virtue of tokening the above-mentioned three sentences, then "presentness is successively possessing different super-properties" can be used to express this awareness.

The maximal expansion of my awareness results in a specific alteration in the quality of my sensuous feelings in addition to their general alteration to a sublime quality. The sublime quality appears when I focus on presentness rather than on this or that first order being, but when I expand my awareness to take into account all of the absolutely infinite number of beings, there is a further modification to my feelings. My emotions are now about presentness in respect to its successive posses-sion of different super-properties, such that what now matters to me is not this state of affairs or that one but all of them equally. Normally what matters to me are certain things or events or their relation to me ("I like this," "That bodes ill for me," etc.) but now my experience of what matters is cosmically present-centric. I experi-ence a total detachment from any selected or highlighted aspect of reality that might ordinarily be distinguished as "being more significant to me now than the other aspects of reality" and experience a total attachment to presentness in respect to its possession of cosmical properties (i.e., super-properties). I lose each being and regain them all in the absolutely infinite, the super-state of affairs, which alone matters. I experience a *cosmical* feeling-quality, which is a sub-type of the generic sublime feeling-quality.

My cosmical attitude *de presenti* may be identified with the vision of the unity of the infinitely extended reality. There is only one super-subject (a subject that pos-sesses super-properties) and everything else is a proper or improper part of the super-subject's super-properties. The whole plane of reality appears in its aspect as a succession of vast absolutely infinite adjectives of presentness.

In the presentist experience I become transfigured. It may be said metaphori-cally that "my mind becomes one with presentness." In literal terms, this means that my conscious experience is such that it would coincide with an omniscient con-sciousness possessed by presentness if presentness were conscious. If presentness were conscious of everything in which it inhered, it would be related by the relation of *consciousness* to everything to which it is non-relationally tied by the property-tie of *inherence*. This "coincidence in extension" of consciousness and inherence would in fact obtain if I were identical with presentness, if my conscious acts were performed by presentness itself. If this were the case, presentness would be con-scious of itself (which would also be my self) as inhering in all beings. To experi-ence this metaphorical "oneness with presentness," my attitude must change from being ego-centric to present-centric, such that I now relate (in the "as if" mode) my consciousness to presentness rather than to myself (considered as distinct from presentness). The point or purpose of this present-centric "transfiguration" of consciousness is to attain maximal self-transcendence, to be able to view all of reality from the perspective of presentness rather than from my perspective, and to thereby achieve the greatest possible self-detachment and attachment to presentness in respect to its inherence in every being. This allows my psychological unity to mirror to the maximal extent the unity of reality. In this respect, I realize one of the human excellences, the excellence of an affective appreciation of the unity of reality as a whole.

This transfiguration of consciousness is most accurately described by saying that I come to see reality from a merger of my perspective with the hypothetical perspective of presentness, since I still experience reality from my maximal indexical location. I grasp presentness as inhering in my surroundings and in each of the absolutely infinite number of beings that are beyond my surroundings. This involves a singling out awareness of my sensory field and a general awareness of the beings that are not sensorily apparent to me. A maximally omniscient consciousness that would be possessed by presentness would coincide exactly with my specific awareness of my sensory field but would also coincide exactly with each other sentient organism's awareness of its sensory field and would in addition include a specific awareness of each other being. Just as presentness *inheres in* every sensory field, just as this field is apparent to the pertinent organism, so presentness would be conscious of every sensory field, just as this field is apparent to the organism. My consciousness coincides exactly only with the consciousness that presentness would possess of *my* sensory field, and coincides only generally with the consciousness that presentness would have of every other field and every other being.

By virtue of this self-transcendence, I become psychologically free of the problems, solutions and concerns of my ego and its normal ego-centric attitude, wherein it ascribes its consciousness of reality to itself alone (and not to presentness). When transfigured, I am, hypothetically, presentness and not (merely) this mortal and limited being. Transfigured, I am free, above the clouds, as the untrammeled consciousness possessed by the universal metaphysical subject.

Index

DATE DUE

5/1/11			